Practicals in Plant Physiology and Biochemistry

About the authors

Dr. Manju Bala has done her doctorate from Punjab Agricultural University, Ludhiana in 1998. She is working as a Senior Scientist (Plant Biochemistry) at Directorate of Rapeseed-Mustard Research (ICAR), Sewar, Bharatpur, Rajasthan. She also has experience of working as Assistant Professor and Vice-Principal at M.N. Institute of Applied Sciences, Bikaner, a college affiliated to MGS University, Bikaner. She has taught various courses on Biochemistry and Biotechnology at undergraduate and post-graduate levels. She has guided 20 M.Sc. students, one Ph.D. student and contributed 10 articles/bulletins, four book chapters and 25 research papers in various National and International Journals.

Dr. Sunita Gupta is working as Associate Professor of Plant Physiology at SK Rajasthan Agricultural University, Bikaner. She has done her Ph.D. from Rajasthan Agricultural University, Bikaner and gained research exposure at University of Adelaide, Australia. She is actively engaged in research and teaching of plant physiology to undergraduate and post graduate students. Her major research interests are photosynthesis, stress physiology and growth regulators. She has published a book on plant physiology, more than 30 research papers in referred journals and guided many post graduate students.

Dr. N.K. Gupta has obtained his Ph.D. from Indian Agricultural Research Institute, New Delhi in 1997. Presently, he is working as Associate Professor of Plant Physiology at SK Rajasthan Agricultural University, Bikaner. He has also worked as visiting scientist at University of Adelaide, Australia. Dr. Gupta is actively pursuing research on genomics of drought, salinity and high temperature tolerance in plants. He has guided many master's and doctoral students for their dissertation. Till date, one text book on plant physiology, three book chapters, five review articles and thirty five research papers are in his credit.

Dr Manjeet Kaur Sangha obtained her Ph.D. from Post Graduate Institute of Medical Education and Research (PGIMER), Chandigarh. She joined Department of Biochemistry, Punjab Agricultural University (PAU), Ludhiana, Punjab, in 1998 as Assistant Biochemist and now working there as Senior Biochemist. She has been engaged in teaching of undergraduate and postgraduate courses. She has guided seven M.Sc. students and one Ph.D. student. She is working on biotic and abiotic stresses in crop plants and associated with projects funded by IFPRI, Harvest Plus, Canada, ICAR and DBT, New Delhi. She has published two book chapters, two review articles and about 25 research papers in various national and international journals.

Practicals in Plant Physiology and Biochemistry

Dr. (Mrs.) Manju Bala
Dr. (Mrs.) Sunita Gupta
Dr. N.K. Gupta
Dr. (Mrs.) Manjeet Kaur Sangha

SCIENTIFIC PUBLISHERS

Published by

SCIENTIFIC PUBLISHERS (INDIA)
5 A, New Pali Road, P.O. Box 91
Jodhpur 342 001 (INDIA)
E-mail: info@scientificpub.com
Website : http://www.scientificpub.com

Print : 2024

ISBN: 978-81-72338-52-7
ISBN: 978-93-87741-99-7 (E-book)

Printed in India

CONTENTS

PART A - PLANT PHYSIOLOGY

Practicals in Plant Physiology and Biochemistry

Dr. (Mrs.) Manju Bala
Dr. (Mrs.) Sunita Gupta
Dr. N.K. Gupta
Dr. (Mrs.) Manjeet Kaur Sangha

SCIENTIFIC PUBLISHERS

Published by

SCIENTIFIC PUBLISHERS (INDIA)
5 A, New Pali Road, P.O. Box 91
Jodhpur 342 001 (INDIA)
E-mail: info@scientificpub.com
Website : http://www.scientificpub.com

Print : 2024

ISBN: 978-81-72338-52-7
ISBN: 978-93-87741-99-7 (E-book)

Printed in India

Preface

Mere theory may not help in understanding the science in a holistic manner and hence there is a need to give emphasis on practicals, along with theory. The aim of this book is to provide fundamental and up-to-date knowledge on the subject of plant physiology and biochemistry. This book will be of use to under–graduate students opting plant physiology and biochemistry as well as to post–graduate students of plant physiology, biochemistry, botany, biotechnology and microbiology.

The text has been divided into two sections: Part A - Plant Physiology, and Part B-Biochemistry. Plant Physiology section covers experiments, mainly on measurement of photosynthetic parameters, water status, stomatal regulation, pigment extraction and cell membranes. Topics on seed germination, methods to break seed dormancy, leaf area, growth analysis, plant nutrition and mineral deficiency have also been covered.

In Biochemistry section, experiments relating to carbohydrates, proteins, lipids and nucleic acids have been discussed. Topics on buffers and enzymes have also been dealt with. A brief study on vitamins and secondary metabolites has been introduced. Techniques of chromatography and electrophoresis have also been covered.

The idea of preparing this document is to create an enduring interest in practical aspects of the subject, which will come a long way in its application. Proper emphasis has been given to meet the demands of the syllabus prescribed by State Agricultural Universities as well as traditional universities.

Utmost care has been taken during the preparation of this book but still there may be some unintended mistakes. Valuable suggestions and comments from readers are always welcome for further improvement.

Dated: January 05, 2013

Dr. Manju Bala
Dr. Sunita Gupta
Dr. N.K. Gupta
Dr. Manjeet Kaur Sangha

CONTENTS

PART B - BIOCHEMISTRY

PART A - PLANT PHYSIOLOGY

1. PREPARATION OF SOLUTIONS

Solution is a homogeneous mixture of two or more substances, in which the molecules or ions of one substance are dispersed throughout the medium of other substance. The substance, which is dispersed, is called solute, and the substance, in which solute particles are dispersed, is called solvent.

The concentration of a solution may be defined as the amount of the solute present in the given quantity of the solvent. The concentration of a solution can be expressed as:

a. Percent Solution

Parts of a solute dissolved to form hundred parts of solution is known as percent or percentage solution. The percentage of a solution can be expressed in three ways:

- Weight per unit weight (w/w) - A 1% w/w solution has 1 g of solute and 99 g of solvent to make 100 g of solution.

- Weight per unit volume (w/v) - A 1% of w/v solution contains 1 g of solute in 100 ml of solution.
- Volume by volume (v/v) - A 1% of v/v solution contains 1 ml of solute in 100 ml of solution.

b. Parts Per Million (ppm) Solution

If solution is very dilute or solute is present in very minute quantities, the concentration is expressed in parts per million instead of per cent. Parts of a solute present in million parts of a solution is known as a ppm solution. If we dissolve 1g of any substance in 10,00,000 g of a solution, it would be 1 ppm solution. In simplified way, 1 ppm means 1 mg dissolved in 1 *l* of a solution. Similarly, parts of a solute present in billion (10^9) parts of solution is known as a ppb solution.

c. Molality (m)

It is defined as the number of moles of a solute dissolved in 1,000 g of the solvent. To make 1 molal solution of NaOH in water, dissolve 40 g of NaOH in 1,000 g of distilled water.

d. Molarity (M)

It is defined as the number of moles of a solute dissolved per litre of the solution.

1 mole = 1g molecular weight; Molarity = Gram molecular weight/litre of solution

The one mole or the gram molecular weight of NaOH is 40 g. If we dissolve 40 g of NaOH into distilled water to make its volume 1*l*, the solution will be called 1 molar (1 M).0.5 M solution of NaOH can be prepared by taking one half of the mole (20 g) in one litre of the solution.

e. Normality (N)

It is defined as the number of gram equivalents of a solute dissolved per litre of the solution.

Equivalent weight = Molecular weight/Acidity or Molecular weight/Basicity

1 N solution of an acid contains one gram equivalent weight of H^+ per litre of solution. Let us prepare 1 N solution of HCl and H_2SO_4. We know that the molecular weight of HCl is 36.5. It is a monobasic acid as it has only one replaceable H^+. The gram equivalent weight of HCl is also 36.5 g. So, weigh 36.5 g and dissolve it in water to make its volume to 1 litre.

H_2SO_4 is a dibasic acid and has two replaceable H^+. Molecular weight of H_2SO_4 is 98 and equivalent weight is 49 (molecular weight/2). So, 49 g of H_2SO_4 when dissolved in 1 litre of solution gives 1 N solution of it.

The concentration of solute per litre can be determined from its normality as below:

Normality = Strength (concentration in gram per litre) / Equivalent weight

Strength (concentration in gram per litre) = Normality x Equivalent weight

EXERCISES

1. Prepare 5% solution of $CuSO_4.2H_2O$.
2. Prepare 5 ppm solution of NaCl.
3. Prepare 1.0 m solution of glucose.
4. Prepare 0.2 M solution of NaCl.
5. Prepare 0.1 N solution of NaOH.

2. DETERMINATION OF WATER STATUS IN DIFFERENT PARTS OF PLANTS

Principle

Relative water content (RWC) is the most dependable and easy technique for measuring the water status of a tissue. RWC expresses the water in the original sample as a percentage of water in the fully hydrated tissue.

Materials

Fresh plant material (e.g. leaves), distilled water, butter paper bags.

Glassware and Equipments

Petriplates, leaf punch, electronic balance, oven.

Procedure

RWC can be measured by the following methods:

2.1. Whole Leaf Method

Select physiologically functional leaves of different crop plants or genotypes. Leaves are the most preferred plant materials since they are physiologically more active. Keep the leaf sample carefully in polythene bags while carrying them to the laboratory to minimize the transpiration losses. Once brought into the laboratory, remove the petiols or sheaths and record fresh weight of the laminae immediately (take a minimum of three replications for each sample). Float the leaf sample in petriplates containing distilled water for 4 h. Blot the leaves carefully to remove water on the surface and record their turgid weight. Dry the sample in an oven at 70°C and record its dry weight. Calculate RWC using the following formula:

RWC = [Fresh weight–dry weight/Turgid weight–dry weight] x 100

2.2. Leaf Disc Method

Make 25 discs of each sample of about 1 cm diameter with a close punch. Record their fresh weight using the electronic balance. Float the disc in petriplates containing distilled water at room temperature for 4 h. Record the turgid weight of the disc after carefully blotting the excess water. Keep the disc in butter paper bags for drying in an oven at 70°C. Record their dry weight and calculate RWC using the formula given in the Whole leaf method.

Precautions

- Selection of plant material (stage of the plant, leaf position, treatment and time of sampling) should be proper.
- Use polythene bags to carry the material to the laboratory.
- Blot the excess water from leaf surface before recording turgid weight.

3. DETERMINATION OF WATER POTENTIAL

Background

Water potential is the difference in free energy or chemical potential per unit molal volume between pure water and water in the cell at the same temperature. The water potential of pure water at atmospheric pressure is equal to zero. Hence, the water potential of water in cells and solution should typically be less than zero or negative.

The water potential is affected by all factors, which change the free energy or chemical activity of water molecules. Thus, water potential is increased by development of turgor (pressure) and by rise in temperature. It is decreased by addition of solutes, matric force that absorbs or binds water, negative pressure (tension), and reduction in temperature. The water potential can be measured by Weight method and by Chardakov's dye method.

3.1. Weight Method

Principle

As a common method of measuring water potential in plant tissue, uniform sample of pieces of tissues are placed in a series of solution of a non–electrolyte such as sucrose or mannitol. The concentration at which the weight and volume of the tissue does not change, indicates neither net loss nor net gain in water. Such a situation would mean that the tissue and the solution are in osmotic equilibrium to begin with and so the water potential of the tissues must be equal to water potential of the external solution.

Materials

Sucrose or mannitol, paper towel, graph sheet.

Glassware and Equipments

Beakers, cork borer, blade, analytical balance.

Procedure

Take beakers (150–250 ml) containing 100 ml of distilled water with 0.0 (control), 0.05, 0.10, 0.20, 0.25, 0.30, 0.35, 0.40, 0.45, 0.50 and 0.60 molal sucrose solutions. Take potatoes and use the cork borer of approximately 1 cm diameter. Obtain 12 cylindrical sections from a single potato; each of atleast 2–4 cm long. Cut all the 12 sections to measured uniform length with a razor blade leaving a cut at the end of each cylinder. Place the cylinders below the folds of a moist filter paper, on which the position of the cylinders are denoted by the series of concentration of sucrose to be used. Weigh each cylinder with analytical balance to the nearest milligram. After each of the cylinders is weighed, cut the cylinder into uniform slices (approximately 2 mm thick) immediately, and place the collective slices from one cylinder in one of the test solutions. After incubation for 1.5–2.0 h, remove the slices comprising each sample, and blot them gently on filter paper and weigh them. The weighing should be done in chronological order, in which they were initially placed. Note down the molal concentration of sucrose solution, at which there is neither loss nor gain in potato tuber weight. Then, construct a graph following the changes in weight or per cent change in weight versus sucrose concentration.

Calculations

Calculate the water potential (Ψ) using the following formula:

$$\Psi = -\,mi\,RT$$

Where,

m = Molality of the solution

i = Ionization constant (1.0 for sucrose)

R = Roul's gas constant (0.083 litre bar/mole degree)

T = Absolute temperature (°C+ 273)

The above formula needs to be used to calculate the Ψ for only one of the sucrose solutions. The Ψ values for all other solutions can be more simply calculated using the following formula:

$$m^1 / \Psi\pi^1 = m^2 / \Psi\pi^2$$

From the graph, determine the sucrose concentration, at which no net change in weight occurred. Calculate the Ψ for this solution. This value equals to the water potential of the tissue.

Precautions

- Obtain all cylindrical pieces from a single potato tuber.
- Use the analytical balance to the nearest milligram.
- Weigh the samples in the chronological order in which they were initially placed in the test solution.

3.2. Chardakov's Dye Method

Principle

Water potential is a diagnostic tool, which tells whether a plant cell would absorb water from a more turgid cell or will supply water to a less turgid cell i.e., it determines the direction of spontaneous water movement and attainment of water equilibrium. Chardakov's dye method is based on the change in density of the solution.

Materials

Sucrose, leaf sample, dye (methylene blue or methyl orange or methyl red).

Glassware and Equipments

Test tubes, dropper, pipette, beaker, measuring cylinders, punch machine.

Procedure

Prepare 1 M stock solution of sucrose. From this stock solution, make graded series of sucrose solution (ranging from 0.1–0.8 molar). Keep this graded sucrose solution in eight separate test tubes and label this series of tubes as 'control series'. Then, set up another series of the same sucrose solution and label it as 'test series'. Place the control series of a solution in a row in order of increasing concentrations. Then, place the test series of tubes in corresponding position in the second row. Stopper the tube until ready for use. With the help of punch machine make the leaf disc of equal size. Transfer the sample immediately in each of the test tubes in the test series.

Add a pinch of methylene blue powder in each tube of control series. After 30 min, withdraw a small amount of solution from the first tube of the control series with a dropper. Then, hold the dropper about one inch under the surface of the solution of the corresponding concentration in the test series and slowly release the drop of blue test solution.

If the drop rises, the solution in test series has increased in density due to uptake of water by the plant tissue. If the drop sinks, it indicates that the solution in test series has been diluted and decreased in density. However, when the density of the test and control solutions is similar, the drop will neither move up nor down, but will diffuse indicating a dynamic equilibrium between sucrose solution and plant tissue.

Perform the above operation at 30 min intervals during the next two hours. Record the observations and determine the water potential of the tissue. Note down the observations as given below.

Table 1. Observations to be recorded to determine the water potential of the tissue

Sucrose solution (M)		Pattern of movement of control solution	Nature of sucrose solution
Control series	Test series		
0.1	0.1	Upward	Hypotonic
0.2	0.2	Upward	Hypotonic
0.3	0.3	Upward	Hypotonic
0.4	0.4	Diffuse	Isotonic
0.5	0.5	Downward	Hypertonic
0.6	0.6	Downward	Hypertonic

Calculations

Observe the point where the drop of the control solution when placed in corresponding test solution showed diffusion. It implies that the water potential of the tissue is equal to the osmotic potential of sucrose solution showing clear diffusion of dye solution. In the present case, the water potential of the tissue is equal to the osmotic potential of 0.4 M sucrose solution, which is equivalent to –11.1 bars.

3.3. Measurement of Water Potential by Pressure Chamber Apparatus

Principle

This method actually measures the xylem potential. Water in the xylem of plant is held with some tension. If the tension is released by cutting of leaf or branch, which breaks the water column, the xylem sap falls back. The pressure applied in the chamber, just to initiate the oozing out of xylem sap from the cut end, is an estimate of the original tension of the xylem prior to excision.

Material

Plant material (fully expanded leaves)

Glassware and Equipments

Pressure chamber (Fig. 1) apparatus, gloves, note book.

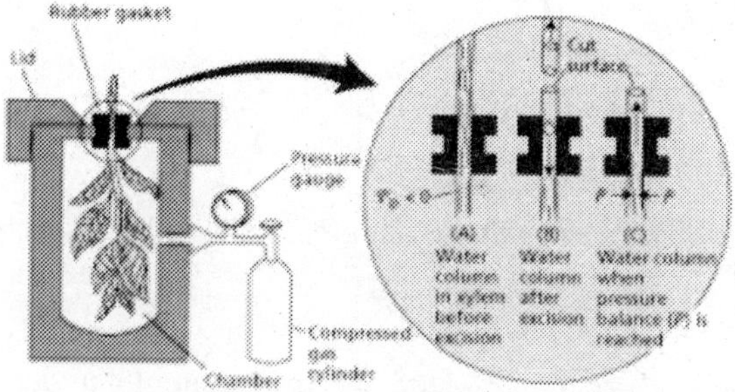

Fig. 1. Pressure chamber

Procedure

A leaf or branch is enclosed in an airtight chamber of the apparatus in such a way that the cut end of petiole or leaf is 1 cm above the chamber lid. The pressure inside the chamber is gradually increased by compressed dry nitrogen gas from a cylinder until xylem sap starts oozing out of the cut end of the leaf. This can be detected with a hand lens or binocular microscope. Record the pressure in pressure gauge attached with the instrument. After releasing the nitrogen gas from the pressure chamber, place another sample in the pressure chamber and measure xylem potential as mentioned earlier. The measurement should be made at a fixed time (and should be recorded).

Results

The pressure applied in the chamber to enable the xylem sap to ooze out from the cut end of the leaf tissue is exactly equal to the original tension of xylem prior to excision.

4. CELL MEMBRANE STABILITY TEST

Principle

The environmental stresses in plants affect cellular membranes, which disturbs its function partly or totally. The cellular membrane dysfunction due to stress can be expressed in increased permeability and leakage of ions, which can be readily measured by the efflux of electrolytes. Hence, the estimation of membrane dysfunction under stress by measuring cellular electrolyte leakage from affected leaf tissue into an aqueous medium is finding a growing use as a measure of cell membrane stability (CMS) and as a screen for stress resistance.

Materials

Leaf samples, double distilled water

Glassware and Equipments

Conductivity meter, water bath, electric balance, test tubes

Procedure

Take 100 mg leaf discs and wash thoroughly under running tap water followed by washing with double distilled water. Keep these leaf discs in test tubes containing 10 ml distilled water and heat at 40°C for 30 min. Measure the electrical conductivity (C_1) by conductivity meter. Subsequently, keep the same samples in a boiling water bath (100°C) for 10 min and take the electrical conductivity (C_2). The CMS can be calculated as:

$$CMS = [1-(C_1/C_2)] \times 100$$

Precautions

- Samples should be thoroughly washed.
- Conductivity meter should be calibrated properly before use.

5. MEASUREMENT OF PHOTOSYNTHETIC RATE USING INFRARED GAS ANALYZER (IRGA)

Background

Measurement of CO_2 uptake provides an alternative and direct method of measuring productivity. It is instantaneous, non destructive, allow separate investigation of individual leaf and measure photosynthetic gain from respiratory losses. Assimilated rate (A) is expressed as the amount of CO_2 assimilated/ unit leaf area/ time ($mol.m^{-2}s^{-1}$). The rate of CO_2 assimilation by the material enclosed is determined by measuring changes in CO_2 Concentration of the air flowing across the chamber.

Principle

Assimilation of CO_2 has been measured by many different techniques like $^{14}CO_2$ labeling, conductivity and IR spectroscopy. Infra Red Gas Analysis of CO_2 is the most wide spread contemporary method of determining photosynthesis and respiration, because of the sensitivity, reliability, accuracy and simplicity of this technique. Heteroatomic gas molecules absorb radiation at specific sub milli-meter infra red wavebands each gas having a characteristics absorption spectrum. Gas molecules consisting of two identical atoms (e.g., O_2, N_2) do not absorb infra red (IR radiation) and do not interfere with determination of the concentration of heteroatomic molecules. IR gas analyzer has been used for measurement of a wide range of heteroatomic molecules CO_2,H_2O, NH_3,CO,N_2O,NO and gaseous hydrocarbon. Major absorption band of CO_2 is at l= 4.25 μm with second peak at l= 2.66, 2.77 and 14.99 μm. Water vapors also absorb at 2.77 μm and can interfere. This problem is overcome by drying the air, or by filtering out all radiations at the wavelengths where absorption by two gases coincides. Absorption of radiation by CO_2 at any one wavelength follow Beer- Lambert's Law and depends upon the radiation path length through the gas and molar concentration of CO_2 (Fig. 2).

Photosynthetic system

Photosynthetic system consists of three basic parts i.e. Infra-red source, gas cell and detector. CO_2 in the gas cell will decrease the radiation reaching the detector, causing a decrease in detector output signal for true differential measurements. Two chambers are needed and detector must be capable of measuring the difference in amount of radiation absorbed in two cells.

IR Source

Two type of IRGA may be recognized, DIR (dispersive) and NDIR (Non dispersive). DIR- Source radiation is passed through a monochromatic and the selected narrow band is passed through the cell. Thus the sample may be scanned over a range of wavelengths and absorption by several heteroatomic species measured. NDIR- Broad band radiation is used only for one molecular species. IR source is typically a spiral of nichrome alloy or tungsten heated on about 600-800°C through a low voltage circuit. The IR source should be well mounted to avoid contamination. The source radiation is chopped either mechanically with a rotating shutter or electronically by pulsing the electric supply or by solid state chopping.

Cells

Most IRGA used for photosynthesis measurements are dual beam instruments passing equal amounts of radiation into two parallel cells analysis (1) continuous flow of CO_2 and (2) reference factory sealed (with CO_2 free air). Cell therefore contains a gas inlet and an outlet, window of an IR transmitting material such as calcium fluoride, and highly reflective inner lining. Inner surface are gold plated to maximize transmission. Path length employed is 250 mm to obtain 1 μmol/mol resolution.

Detector

Most common detector – luft type, operates on the principle of positive filtration-absorb IR in the CO_2 absorption band. This is achieved by filling the detector with CO_2. The detector is divided into two chambers separated by a thin diaphragm of copper-beryllium, aluminum or gold. In parallel configuration, radiation passing through the reference cell enters one chamber and radiation passing through the analysis cell enters another. Both chambers will absorb radiation in the CO_2 absorption band, the amount of radiation available for absorption being inversely proportional to the amount absorbed in the cell. The chopped radiation will thus cause periodic pressure changes in the detector with simultaneous vibration of the membrane. The amplitude of the vibration is determined by the pressure difference between analysis and reference cell change in amplitude produces a change in the condenser capacity inversely proportional to voltage change across the condenser.

Calibration

Although IRGA is highly sensitive to continuously monitoring of CO_2 concentration, they require frequent calibration. Calibration is a process of adjusting the IRGA output either an electric signal or a meter deflection to the quantities and unit required.

Operation

After calibration the leaf is enclosed in a chamber and net exchange of CO_2 between leaf and atmosphere is measured. Monitor the rate at which CO_2 concentration change over short time interval (typically 10-20 seconds). Calculate the photosynthetic rate using a rate of change in CO_2 concentration, area of leaf enclosed and temperature.

Advantages

(1) It is a portable instrument that makes rapid and simultaneous measurement of photosynthetic rate and stomatal conductance with minimal change to the environment of the plant material.

(2) With typical CO_2 depletion of only 2 to 10 ppm (depending on the photosynthetic rate) and very small change in RH, the measurement is made very near to ambient conditions.

(3) Since the chambers of different types are available so it is possible to measure the photosynthetic rate ranging from small leaflets to whole plant or canopy measurement.

Precautions

1. The green area of leaf should cover the complete area of cuvette.
2. Calibration should be done regularly.

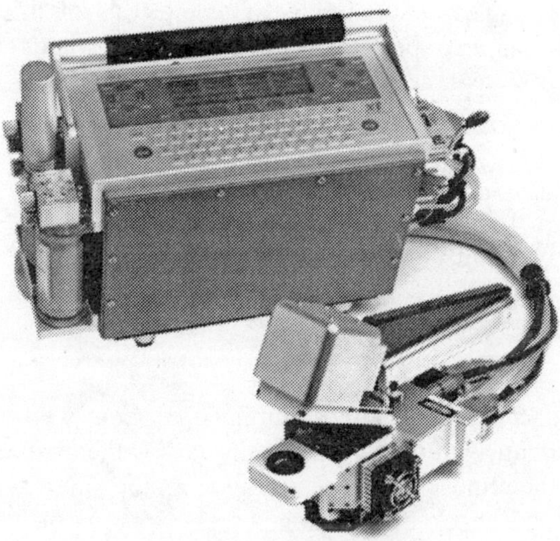

Fig. 2. Infrared gas analyser

6. MEASUREMENT OF TRANSPIRATION

Background

Transpiration is the loss of water in the form of vapours from the aerial part of the plant. The greater part of the water loss from the plant tissue occurs through small openings present on leaves known as stomata, and this type of transpiration is called stomatal transpiration. 80–90% water loss in plants occurs through stomatal transpiration. Some loss of water also occurs through cuticles present on the surface of leaves and herbaceous stems; this is called as cuticular transpiration. The cuticular water loss occurs about 10–20%. In addition to stomatal and cuticular transpiration, a small amount of water loss also occurs through lenticels, which are small openings in corky tissues covering stem and twigs; this is called lenticular transpiration. The magnitude of water loss through lenticels is insignificant (0.1–1%). The rate at which plant and plant parts lose water by transpiration is naturally of interest to plant scientists.

6.1. Gravimetric Method

Principle

Plant weight is measured at definite intervals of time. The water loss per unit leaf area per unit time is estimated and calculated for measurement of transpiration.

Materials

Potted plants, polythene sheet, gloves, blotting papers

Glassware and Equipments

Platform balance, forceps, big size petriplates

Procedure

Select intact plants of almost similar size from well–watered pots. Seal the pot along with root system using a polythene sheet so that the water loss to the atmosphere occurs only from the shoot system and not from the soil (to check evaporation losses). Record the weight of the potted plant immediately and keep them in different environment such as (1) warm environment, (2) cool environment, and (3) calm environment.

Record the weight of the entire pot periodically at one hour interval over specified period (final weight) for each environment. The loss in weight represents the quantity of water transpired. Express the rate of transpiration in grams as water vapour loss per plant per hour.

Calculations

Determine the amount of water loss by subtracting the final weight from the initial weight and calculate the amount of average water loss per plant at the end of the experiment. This will calculate the water–vapour loss per plant per hour. The results can be represented as $g.H_2O\ h^{-1}$ $plant^{-1}$ under different environments.

6.2. Cobalt Chloride Method

Principle

The cobalt chloride paper is dark blue in colour when dry but turns pink when it comes in contact with moisture. Standardization of cobalt chloride paper is done by measuring the time required for the paper to change in colour from blue to pink. The rate of colour change is a measure of the rate, at which moisture is being lost from the leaf.

Materials

3% Cobalt chloride solution, filter paper, vaseline, healthy potted plant

Glassware and Equipments

Oven, glass slides, clamps

Procedure

Soak few pieces of filter paper (1 cm^2) in 3% slightly acidic solution of cobalt chloride and dry them in an oven at 50°C. After that keep them in desicator. The property of cobalt chloride paper is that they are deep blue when dry but in contact with moisture may turn pick. Standardise the cobalt chloride paper by measuring the length of time required for the paper to change colour from blue to pink by placing on wire screen a few mm above the surface of liquid water on petridish.

After standardising the cobalt chloride paper take two dried cobalt chloride papers, each 1 cm^2. Place one paper on the upper surface of the leaf of a potted plant and the other on the lower surface. Cover both the filter papers with glass slides and clamp them properly to the leaf by a metal clip. To check the water losses from the edges of the leaves, seal the leaf with vaseline quickly. As the water lost in transpiration from the leaf is absorbed by the filter paper, it turns pink in colour. Observe the change in colour of both the filter papers. The rate of colour change is a measure of the rate, at which moisture is being lost from the leaf. This provides a qualitative measurement of the water–vapour loss from the leaf surfaces.

Calculations

Calculate the relative rates of water–vapour loss from the upper and lower epidermis of the leaf by calculating the transpiration index:

Transpiration index = Time for standard colour change (in s) / Time for the colour change on leaf (in s). Higher the resulting ratio, faster will be the rate of transpiration.

Results

The time taken to change the colour from blue to pink is less in lower epidermis than in upper epidermis. It indicates high rate of transpiration in lower epidermis because of greater number of stomata on lower epidermis than on upper epidermis. Also interpret your result with respect to the relative rate of transpiration from both the surfaces.

For quantitative measurement calculate water vapor loss (G) in terms of $g/dm^2/h$ for upper and lower surface of each leaf by the given equation:

$$G = \frac{3600\,x}{T}$$

x = Number of grams of water vapour required to change 1 cm^2 of cobalt chloride paper from blue to pink. This is computed by recording the initial and final weight of cobalt chloride paper.

T = Observed time of colour change in seconds.

3600 = Conversion factor from unit per second to hour.

6.3. Ganong's Potometer

Principle

Water is taken up by a leafy twig, connected to the apparatus. As water is taken up, a small air bubble is drawn along a horizontal capillary tube, so that the rate of movement of the bubble corresponds to the rate of water uptake by the shoot.

Materials

Paraffin wax, twig containing leaves, eosin, water

Glassware and Equipments

Ganong's potometer, beaker

Procedure

The Ganong's potometer consists of a horizontal graduated tube. One end of the horizontal tube is bent upward, which is wide open, and the other end is long graduated, which is bent downward. A reservoir is fixed to the horizontal tube near the wide open end. There is a stopcock just below the water reservoir to control the water flow. The complete apparatus is made airtight by paraffin wax and fixed on a wooden stand.

Fill the apparatus with water through water reservoir by opening the stopcock connected with the water reservoir. Cut a twig or a small plant under water and fix it to the wider arm through a split cork. Dip the other bent end of the graduated tube into a beaker containing water coloured with eosin.

To start the experiment, introduce an air bubble in the horizontal tube by lifting the tubes for a moment and dip into the water again. Keep the apparatus in bright light and observe the movement of air bubble. As the transpiration takes place, air bubble moves in the tube from one end to the other end. After it has reached the reservoir, open it to push the air bubble back and then observe its movement again. The time taken for the bubble to move from one end to another end represents the relative rate of transpiration (Fig. 3).

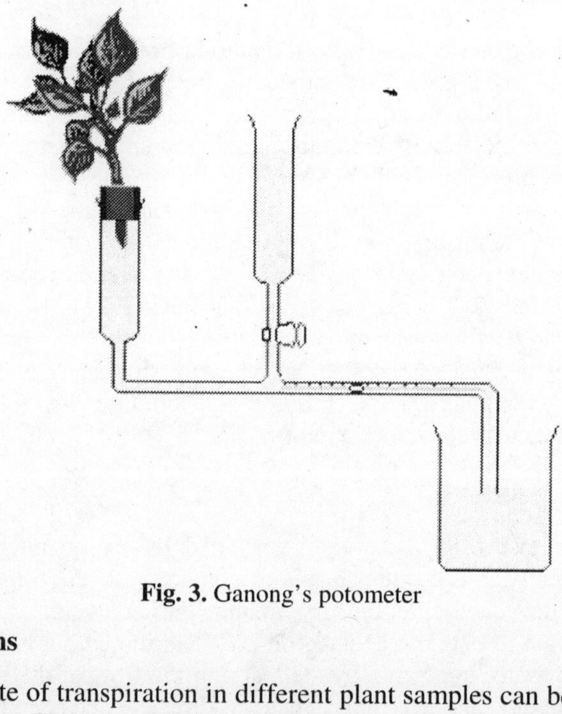

Fig. 3. Ganong's potometer

Calculations

The rate of transpiration in different plant samples can be calculated by calculating Volume of water absorbed or transpired by the sample per second = Volume of water absorbed or transpired by the plant sample / Time

The volume of water can be calculated by the formula:

$$V = \pi r^2 l$$

Where,

V = Volume (ml) of water

π = 22/7

r = Radius of the graduated tube in mm.

l = Distance travelled by the bubble

ml ≈ mg

So, Transpiration rate = mg/sec (By whole twig)

We can also compute transpiration rate per unit leaf area per second by measuring the total area of the plant sample.

6.4. Steady State Porometer

Background

The porometer is an advanced method of measuring transpiration. It has certain advantages over other methods. It is portable and the calibrations are eliminated. The measurements are rapid and there is no need to shape the leaves. Ventilation radiation shields and maintains the temperature of cuvette within 3°C of ambient. The leaf temperature is directly measured through the thermocouple that comes in contact with the leaf, and the air and vapours are circulated by a fan in the cuvette; hence it creates a constant boundary layer. The data can be recorded and can be transferred to a cassette type. The Steady State porometer has a set of standards for measurement of stomatal resistance and has wide applications in plant sciences. It can measure stomatal conductance (cm S^{-1}), leaf temperature (°C), rate of transpiration (μg $cm^{-1}S^{-1}$), and photosynthetically active radiation (μE cm^{-2} S^{-1}).

Principle

Water loss from a leaf is determined by maintaining a constant vapour directly in a cuvette that is in contact with the transpiring leaf. This is achieved by pumping dry air into the cuvette at an appropriate measured rate to obtain a balance (at pre–determined humidity) between the flow of water transpired by the leaf and the flow of the moist air out of the cuvette. Stomatal resistance is directly determined from the measured parameter (Fig. 4).

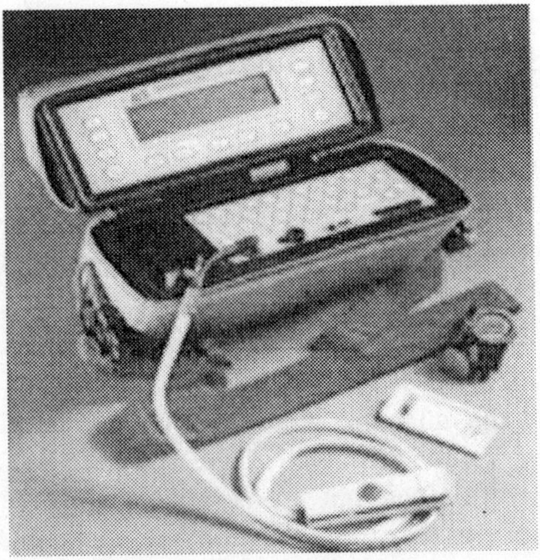

Fig. 4. Steady state Porometer

Procedure

1. Pre–operation Procedure

Prior to making arrangement with the Steady State porometer (LI-1600), the following operations should be performed on a daily basis:

a. Refill the console desiccant pack with freshly dried desiccant. Sensor head desiccant should be changed weekly otherwise the diffusive resistance may be lower than expected (Silica gel desiccant should be dried for one hour in an open pan at 175°C).

b. Remove the dust and pollen from the cuvette otherwise the cuvette will absorb the water vapour resulting in large volume of diffusive resistance.

c. Connect the sensor head cable and air hose to the system and turn the power switch 'ON'. Move the switch to cuvette temperature. Check the response by touching the thermistor bead by the finger tip.

d. Install the aperture cap and leaf clamp, and enter 'APER SET' and press 'SET'.

e. Acclimate the instrument to ambient condition to operate for 30 min. Keep the null adjust value closed and sensor head leaf clamp opened.

f. Check the RH sensor calibration with a reference measurement of relative humidity (e.g., Aspired psychrometer or Dew point hygrometer). If the displayed RH and temperature–corrected RH is different from the reference RH by 3% or more, recalibrate the RH sensor using 1600 SC salt cup humidity test kit or other humidity reference standard.

g. Inspect the leaf temperature thermocouple for proper adjustment. The thermocouple should be located at the peak of the wire. Check the response of the thermocouple (leaf temperature) by gently touching the bead with the finger tip. 'READOUT' should be the same. If it is different, adjust the Thermocouple psychrometer.

h. Check the quantum response by alternatively covering and uncovering the diffuser. It should be near 2000 m moles S^{-1} m^{-2} under full sunlight (solar noon). It is dependent on the atmospheric condition, location, latitude, elevation and time of the year.

i. Check the dry air leakage. Lower DR value than expected can result from the leak in the air flow system.

j. Every year the instrument should be calibrated for mass flow meter, humidity sensor, temperature sensor, quantum sensor and complete circulating check.

2. Summary of Operation

a. Acclimate the instrument to ambient condition.
   - i. Turn the power switch 'ON'.
   - ii. Close the null adjust valve.
   - iii. Operate for 30 min with sensor head leaf clamp opened and shaded.

b. Set the null point humidity to ambient RH.
   - i. Slowly increase the humidity inside the cuvette by clamping a leaf over the aperture cap.
   - ii. When desired humidity level is reached, press 'HUM SET' for about 1second to select that value as the null point.

c. Select the flow of dry air into the cuvette to select a steady state condition.
   - i. Clamp a representative leaf into the cuvette. If the leaf is transpiring, the cuvette humidity will rise.

ii. Open the null adjust valve to provide a flow of dry air into the cuvette. Continue to adjust this value until the null adjust meter needle is on scale.

iii. When the needle centres and is stable, a steady state condition is in effect at the null point humidity.

d. Start taking measurements on experimental leaves.

i. Generally 10–20 s are needed to achieve steady state condition on leaf with a diffusive resistance less than 5 s cm^{-1}.

ii. Monitor relative humidity on the display when the null point RH is reached. Immediately, move the 'READ OUT' function switch to Differential Resistance and note whether the resistance values are stable. If the values are stable, move the 'HOLD' switch to the leaf and read the data.

3. Interpretation of the Transpiration Data

Be careful about how the transpiration measurement is used. Comparing one leaf with another in a qualitative sense is acceptable but, avoid making estimate of whole water use based upon measured rate of individual leaf. To interpret the real value, use the following formula:

$$Tr = Tc\ (Rs+Rb)/(Rs+Ra)$$

Where,

Tr = Real transpiration (outside the chamber)

Tc = Calculated transpiration (inside the chamber)

Rs = Computed stomatal resistance

Rb = Boundary layer resistance (0.15 s/cm is used in the equation)

Ra = Estimated boundary layer resistance (1.8L/V in Scm^{-1})

4. Operation and Maintenance of Humidity Sensor

a. Do not touch or scratch the surface of the sensor.

b. Do not store below 10% RH or above 90% RH.

c. Do not store in a confined environment with desiccant that may vaporize.

d. While installing a new sensor, install with a surface having a thin film on it towards the cuvette wall. This is the surface with the two leads soldered to it.

e. Clean only with dust free air or rinse with the distilled water if rinsing with water. Before making measurements, allow the sensor to re–equilibrate for one day at appropriate humidity at which you want to use.

f. Do not expose the sensor to solvents, cigar smoke, and sulphur pollutants, as it affects polymer coating on the sensor and may cause failure.

Precautions

- The cuvette temperature may increase (2–3°C) after prolonged use in the direct Sun. As much as possible keep the sensor head shaded between measurements.
- Try to work facing the Sun to prevent shading of leaf before or during the measurements.
- Maintain the natural orientation of the leaf while it is being measured.
- Sometimes the process of making a measurement can cause stomatal closure due to physical agitation of the leaf or sudden exposure to a low boundary layer resistance. Water stressed plants and also the plants grown under controlled environments can often show such effects.
- Set the null point humidity slightly above the ambient RH. It will allow a steady state condition to be achieved more quickly if the null point is set to ambient RH. It reduces the measurement time, which reduces the chance of the enclosure affecting the stomatal resistance.

7. DETERMINATION OF STOMATAL INDEX AND STOMATAL FREQUENCY IN LEAVES

Principle

The numbers of stomata differ on the upper and lower surface of the leaf and also differ in different plant species. It is necessary to measure the stomatal frequency and if we can prepare a replica (an exact copy) or print of stomata using any fluid, we can measure the stomatal frequency or index by counting the stomata in a given leaf area.

Materials

Fluid or nail polish, brush, adhesive celluloid tape, leaf samples, glue

Glassware and Equipments

Glass slide, microscope, forceps

Procedure

We can measure the stomatal frequency by the 'Replica method'. Apply the transparent nail polish or correcting fluid uniformly on either side of a leaf with a brush. After drying, remove the replica gently and mount it in glycerine. Keep the replica on clean glass slide. Count the number of stomata (S) and epidermal cells (E) in a specific area under the microscope. Calculate the stomatal index (SI).

$$SI\ \% = S/S+E \times 100$$

By means of Stage micrometer, the diameter of field of view of the microscope was measured in sq. mm area. Count the number of stomata per unit area and epidermal cells for calculating stomatal frequency and stomatal index.

Precautions

- Epidermal strip should be removed carefully.
- The microscope must be accurate and proper calibration should be done before use.

8. EFFECT OF PLANT HORMONES ON RATE OF TRANSPIRATION

Background

Among many factors degree of stomatal opening and closure plays an important role to affect the rate of transpiration. The opening and closure of stomata is determined by the change in turgor pressure in guard cells of stomata. The turgor of guard cells depends upon the uptake or release of potassium ions by guard cells. Uptake of potassium ions leads to the increase in osmotic pressure and subsequently to the increase in turgor of the guard cells; whereas the release of potassium ions leads to the decrease of osmotic potential and thus the turgor of the guard cells also decreases. Plant hormones such as absisic acid (ABA) and cytokinin regulate the uptake or release of potassium ions in guard cells, thus affecting the degree of opening of stomata and subsequently the rate of transpiration. The present experiment, deals with the effect of ABA and cytokinin on the rate of transpiration in excised wheat leaves.

Materials

10–14 days old wheat seedlings grown under normal laboratory conditions, kinetin solution (10 μM, 3 μM, 1 μM), ABA solution (10 μM, 3 μM, 1 μM), parafilm, marking pen, light sources

Glassware and Equipments

Glass vials (approximately 10 ml size), beaker, razor blades, 1 ml graduated pipettes

Procedure

Prepare seven vials containing 5 ml of the solution as indicated below:

Vial	Solution
1	Distilled water
2	1 μM abscisic acid
3	3 μM abscisic acid
4	10 μM abscisic acid
5	1 μM kinetin
6	3 μM kinetin
7	10 μM kinetin

Select 14 wheat seedlings with leaves of uniform size. Cut off the oldest leaf (at least 7 cm long) from each seedling under water so that air does not enter the conducting tissue. Place two leaves in each vial with their cut ends immersed in the solution. Use squares of parafilm to cover the neck of each vial. Cut two slits in each parafilm and cover the mouth of each vial so that slits hold the leaves in place and form a tight seal around each leaf and the neck of the vial. Make sure that the leaves reach the bottom of the vial. Mark water level in each vial by using a fine marking pen. Place the vials in a beaker and place them under continuous light for 24 h. After 24 h puncture the parafilm and use a graduated 0.1 ml pipette to measure the amount of water that has to be added to each vial to bring the water level back to the original level. Make at least four replicates of each concentration.

Results

Find out the effect of different concentration of hormone on the rate of transpiration. The loss of water in different vials is taken as a measure of rate of transpiration.

9. ESTIMATION OF CHLOROPHYLL CONTENT IN PLANT TISSUES

Principle

Chlorophylls, the green pigment of plants, are the most important pigments active in photosynthetic process. Chlorophyll a and b are the most abundant in higher plants. Both chlorophyll a and b have a cyclic tetrapyrrolic ring structure (porphyrin) with an isocyclic ring containing magnesium atom at its centre. The phytol chain of the chlorophyll extends from one of the pyrrole rings. The chlorophyll a has methyl group (-CH_3) attached to the third carbon and chlorophyll b has an aldehyde (HC=O) attached to the third carbon. In addition to minor differences in the structure, chlorophyll a and b exhibit different absorption spectra. An absorption spectra is the measure of the extent to which a given substance absorbs the light of different colour and wavelength. It is a function of relationship between absorption and optical density (OD) and wavelength expressed in nanometres (nm).With the help of spectronic–20 both OD and percentage transmission can be determined. The OD and percentage transmissions are inversely proportional to each other.

Materials

Fresh leaf material, 80% acetone, distilled water, filter paper, glass markers

Glassware and Equipments

Beaker, funnel, measuring cylinders, balance, spectrophotometer, mortar, pestle

Procedure

500 mg of leaf material is taken in a mortar and grind to make a fine pulp. Add 40 ml of 80% acetone to this pulp. Extract this solution again and then filter it through Whatman filter paper no. 1. Wash the extract

repeatedly with 30 ml acetone first and then with 20 ml 80% acetone. Using a measuring cylinder make the final volume to 100 ml. Then, measure the optical density at 645 and 663 nm on spectrophotometer.

Calculations

The spectrophotometer is calibrated by using 80% acetone as blank sample. The optical density of chlorophyll pigment extracted is determined and then amount of Chlorophyll a and chlorophyll b is determined using Arnon's formulae.

$$\text{Chlorophyll a} = [12.7(A_{663}) - 2.69(A_{645})] \times V/1000 \times W$$

$$\text{Chlorophyll b} = [22.9(A_{645}) - 4.68(A_{663})] \times V/1000 \times W$$

$$\text{Total chlorophyll} = [20.2\ (A_{645} + 8.02(A_{663})] \times V/1000 \times W$$

A= Optical density

V= Final volume of 80% acetone (in ml)

W= Weight of leaf material (in g)

Precautions

- At least 2–3 washings of the pulp with acetone is a must.
- Spectrophotometer should be first adjusted using 80% acetone as blank.
- Two different OD values can be taken to avoid error.
- Exposure to light results in rapid destruction of chlorophyll. So extraction should be done in darkness and measurement made as soon as possible.

10. DETERMINATION OF ABSORPTION SPECTRUM OF CHLOROPHYLL

Principle

Since solutions of pure substances do not absorb the energy of all wavelengths of light equally, a substance may be identified by the unique pattern of wavelengths absorbed. The chlorophylls in plants absorb strongly in the blue wavelengths (about 450 nm) and red wavelengths (about 650 nm) but reflect the green wavelengths (about 525 nm). A plot of absorbance versus visible wavelengths (400–700 nm) for a solution of chlorophyll a shows two major peaks, one at 450 nm and one at 650 nm, and a valley from 500–625 nm. This spectrum is characteristic for chlorophyll a and may be used as an aid in its identification.

Materials

Fresh leaf material, ethanol, distilled water, filter paper, glass markers

Glassware and Equipments

Beaker, funnel, measuring cylinders, balance, spectrophotometer, mortar, pestle

Procedure

Extract chlorophyll from leaves of any plant, e.g., spinach. Shred a medium–sized leaf of spinach and mix with 10 ml of ethanol in a 50 ml beaker. Approximately after 5 min, the solution should have a green–coloured tint. If the solution looks yellow in colour, the extraction is not complete. Filter the mixture of leaves and ethanol into another 50 ml beaker using a funnel with a filter. Discard the residue. For the spectrum of chlorophylls, fill one cuvette with ethanol (the blank) and another with the solution of chlorophylls (the sample). First, check the absorbance at 440 nm. The reading, where both chlorophylls absorb to some extent, should be somewhere between 0.3 and 0.8. If the solution is too concentrated, pour the sample into a graduated cylinder and add an equal amount of alcohol to dilute the solution by half. Check the absorbance at 440 nm again. Dilute further if needed. Record the dilution needed.

Set the wavelength to 400 nm and measure the per cent transmittance (%T) of the sample. Continue to measure the %T every 10 nm between 400 nm to 700 nm, remembering to adjust the instrument to full scale at each new wavelength. Take readings for every 5 nm between 420 nm to 480 nm to make it easier to observe the absorption due to both the chlorophylls. Convert %T to A. Plot absorbance versus wavelength by drawing a smooth line through the points (do not connect the dots).

By measuring the absorbance of an uncharacterized solution over a range of wavelengths and plotting the absorbance value on the Y–axis and the wavelength on the X–axis, one can determine the absorption spectrum of a sample. The absorption maximum of any pure substance in solution is the wavelength where absorption is the greatest.

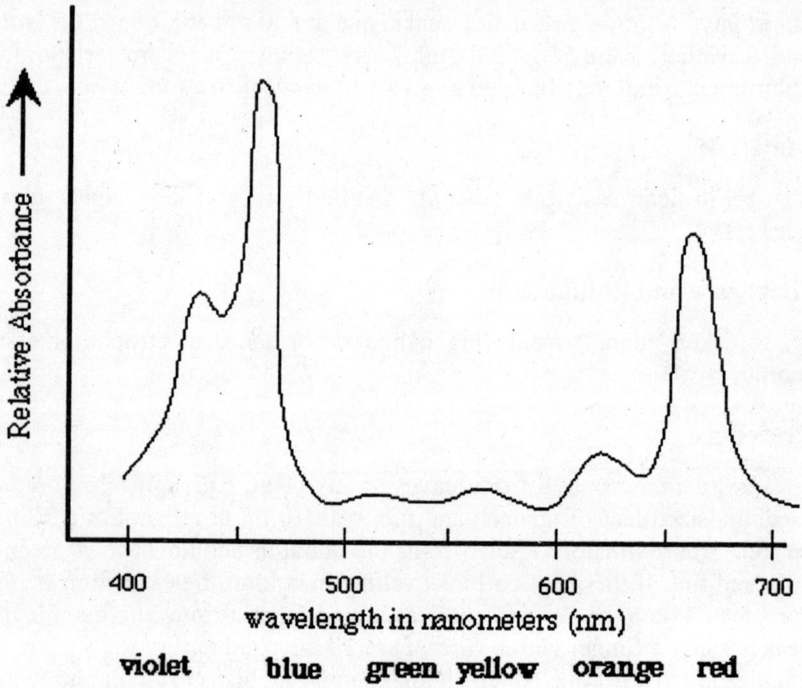

Fig. 5. A graph of absorbance versus visible wavelengths (400–700 nm) for a solution of chlorophyll a

Results

Compare the spectrum with Fig. 5 to find evidence for presence of both chlorophyll a and b. Estimate the relative amounts of chlorophyll a and b in given sample.

11. EFFECT OF CYTOKININS ON CHLOROPHYLL RETENTION

Background

Cytokinins are the plant hormones which are important for chlorophyll retension (greenness) in plants. When leaves are detached from plants, the nucleic acid and protein declines, causing an irreversible yellowing due to the loss of chlorophyll. This process is known as 'senescence'. The progressive and deteriorative processes of senescence can be delayed by treating the detached leaves with cytokinins. Thus, the excised leaves treated with cytokinins remain green for longer period depending upon hormone concentration.

Materials

10–15 days old wheat seedlings, kinetin solution (100, 10, 1, 0.1, 0.01, and 0 μM), 80% ethanol, water

Glassware and Equipments

Petri dishes, Buchner funnel, spectrophotometer, pipettes

Procedure

Grow wheat seedlings in a flat bed of moist soil or vermiculite until the plants reach at a height of 10 cm. Label six petri dishes and pipette 20 ml of kinetin solution into each petri dish. Select leaves of uniform size and excise the section of each leaf between 2 cm and 6 cm from the tip of the leaf. Float five such leaf sections (lower surface down) on the solution in each petri dish. Select additional five leaf sections for the determination of initial amount of chlorophyll in leaf sections. Place petri dishes in plastic bags to prevent evaporation and store them in dark for 4–7 days.

Determine the initial chlorophyll content of five leaf sections by cutting the leaf into small pieces and place them in 15 ml of 80% ethanol in a test tube. Heat the tube in a boiling water bath for 2–5 min to extract chlorophyll. Cool the tubes in dark and adjust the volume of solution to 15 ml by adding 80% ethanol. Read absorbance of the solution at 663 nm and 645 nm on a spectrophotometer. If the absorbance of the solution is over 0.8, dilute the solution with a known amount of 80% ethanol and read the absorbance of the diluted solution. Use this value to calculate the absorbance of the original solution.

Determine the chlorophyll content after 4–7 days in leaf sections treated with different concentration of kinetin in the above described manner. Express the chlorophyll content of leaf sections treated with the hormone assuming the initial chlorophyll content (absorbance) of leaf section as 100%.

Results

Plot a graph of relative absorbance, as a function of chlorophyll content of leaf sections, against the logarithm of kinetin concentrations.

12. SEPARATION OF CHLOROPLAST PIGMENTS

Background

Chlorophyll and carotenoids are the main chloroplast pigments. There are different types of chlorophyll, which differ from each other in their molecular structure and the absorption properties. Chlorophyll a and chlorophyll b are the most abundant pigments in vascular plants. Chlorophyll a is a large molecule with a tetrapyrrole ring and a magnesium atom at its centre. Attached to one of the ring is a long phytol chain (a 20 carbon long insoluble hydrocarbon chain). Chlorophyll b differs from chlorophyll a in having a –CHO group instead of CH_3 group on the pyrrole ring 3. Another class of plant pigment is carotenoids. Carotenoids are 40 carbon compounds. They are red, orange and yellow in colour and fat soluble. Carotene and xanthophylls are two different types of carotenoids. Xanthophyll also contains oxygen in addition to carbon and hydrogen.

Chloroplast pigments can be conveniently extracted from green plant tissue by paper chromatography and thin layer chromatography (TLC).

In paper chromatography, substances placed on one end of the chromatographic paper get deposited on various zones of the paper when an appropriate solvent runs over to the other end of the paper. The mobility of substances on the paper depends on the degree of solubility in the solvent system (mobile phase) and the affinity to the chromatographic paper (stationary phase), which are mostly made up of pure cellulose fibre. Flat paper sheet or round paper cylinders may be used for the separation of substances. Paper chromatography can be distinguished into two types - the ascending chromatography and the descending chromatography. In ascending chromatography, the solvent is placed at the bottom and the paper is hung in such a way that the lower end of the paper is immersed in the solvent. The mixture of substances is loaded at about 2 cm high from the base and the solvent moves up against the gravitational force. In descending chromatography, the solvent is placed on the upper side and the substances are loaded near the upper end of the paper, which is dipped in the solvent. The solvent migrates down the paper by gravitational pull. Due to the gravitational pull, descending chromatography is faster than ascending one.

Thin layer chromatography (TLC) is a relatively new technique for the separation of plant pigments. In this method, a thin layer of an adsorbent is coated on a glass plate or rod, and the sample is spotted on this coating and developed in a tank containing small amount of solvent. These days glass plates or thin films, pre–coated with a large variety of adsorbents, are available commercially. The significant advantage of this method over other methods lie in the fact that TLC method is much faster and requires very little amount of sample. It is much more sensitive than other methods and separation is noticeably much sharper. As long as there is a difference in the adsorption properties of the fraction involved, this method can be successfully used to separate any compound by choosing the right combination of adsorbent and solvent.

The compounds on the chromatogram can be identified on the basis of their diagnostic feature, the 'ratio of fronts' (Rf) values. Rf is the ratio of the distance travelled by the substance to the distance travelled by the solvent in a chromatogram.

$$Rf = \frac{\text{Distance travelled by the substance}}{\text{Distance travelled by the solvent}}$$

12.1. Separation by Paper Chromatography

Materials

Fresh spinach leaves, 80% acetone (v/v), carbon tetrachloride, petroleum ether, benzene, filter paper, chromatography paper, ruler, paper clips

Glassware and Equipment

Mortar and pestle, Buchner funnel, chromatography jar, support rods, glass tubings drawn to a fine tubing, hair dryer

Procedure

Pigment extraction: Grind 500 mg of small cut pieces of fresh leaf material in 20 ml of 80% acetone for about 5 min in a clean mortar. Carefully transfer the resulting green–coloured liquid to a Buchner funnel containing a layer of Whatman no.1 filter paper. Filter the extract using suction.

Separation of pigments: Pour carbon tetrachloride or a mixture of petroleum ether and benzene (9:1,v/v) in a chromatographic jar to a depth of about 2 cm. Allow the internal atmosphere in the jar to equilibrate for few hours. Cut chromatographic paper (Whatman no.1) to a desired size and draw a pencil line about 2 cm away from the bottom. With the help of a glass tubing drawn to a fine tip spot two or three points, about 3 cm apart from one another, with pigment extract. Allow each pigment drop to dry completely before applying the next drops. Drying may be hastened by using a hair dryer. Repeat the application of drops until the marks are dark green. Hang the paper in the chromatographic jar with the lower end dipped in the solvent. Close the jar. Remove the paper when the solvent has moved up to the top of the paper. Allow the paper to dry.

Results

Observe the separation of pigments on the paper. The pigments are arranged in the following sequence from top (solvent front) to bottom - Orange–yellow carotenes, one or more yellow band of xanthophylls, blue–green chlorophyll a, and yellow–green chlorophyll b.

Mark the spots with a pencil since the colours fade away quickly. Calculate the Rf value of each pigment.

12.2. Separation by Thin Layer Chromatography

Materials

Fresh spinach leaves, silica gel G, chloroform, hexane, diethyl ether, petroleum ether, methanol, n–propanol

Glassware and Equipments

Erlenmeyer flasks, Buchner funnel, homogenizer, rotary evaporator, glass plates, gel applicator, mounting boards, desiccator, fan, developing chamber, spectrophotometer

Procedure

Pigment extraction: Extract chloroplast pigment from spinach leaves by grinding approximately 25 g of fresh leaves in 40 ml of acetone in a homogenizer. When the acetone becomes dark green in colour, filter the preparation using Buchner funnel. Evaporate the acetone extract in a rotary evaporator and dissolve the dried pigments in 10 ml of chloroform.

Separation of pigments: Prepare slurry of silica gel G by vigorous shaking of 30 g silica gel with 60 ml distilled water in a stoppered 250 ml or 500 ml Erlenmeyer flask. The slurry must be prepared quickly and used immediately. Using the special mounting board and the applicator provided, coat a number of 5x20 cm glass plates with the slurry of silica gel. The applicator should be set to apply a layer of adsorbent 250 μ thick. Place the coated plates in a drying rack and store them in a desiccator till further use.

Separation: Select a 5 x 20 cm plate coated with adsorbent and run a sheet of tissue paper around the margins to remove excess of flaking gel. On the end of the plate, to which sample is to be applied, make some scratches on both sides at a distance of 1.5 cm from the end. Then scratch a small mark at a distance of 15 cm on each side of the plate above the original marks. The two sets of marks described above mark the origin and the positions to which the solvent will be allow to migrate, respectively. Next, using a 0.1 ml pipette or a very fine capillary tube, apply a small aliquot of the extract between the two marks at the origin end of the plate. Be sure that the sample is spotted near the centre of the origin end of the plate. Avoid touching the pipette to the absorbent and keep the sample spot 1 cm or smaller in diameter. A cool air fan may be used to hasten the evaporation of the solvent if repeated application of dilute extract is necessary. When the sample spot is completely dry, place the plate in a chromatographic chamber containing either one of the following solvents:

- Hexane : Diethylether : Acetone (60:30:20, v/v/v)
- Petroleum ether : Methanol (98:2, v/v)
- Petrolium ether : n–Propanol (99.5:0.5, v/v)

Adjust the depth of the solvent so that the sample spot does not dip into the solvent. Observe the differential migration of various pigments up the chromatogram. When the solvent front has advanced 15 cm above the origin, remove the chromatogram from the developing tank, and allow it to dry.

Results

Make a scaled diagram of the chromatogram, showing all visible pigments. Identify as many of the pigments as possible. Calculate the Rf value from the diagram. When the inspection of chromatogram and recording of all the data is done, remove bands of silica gel containing the individual pigments using a clean razor blade. Place each sample of silica gel containing a single pigment in a centrifuge tube and add 5 ml of acetone. Agitate the tube to facilitate the elution of pigment off the gel. Centrifuge the tube to sediment the gel. Then pipette the supernatant from each tube and transfer the pigment solution from each tube. Adjust the concentration of each pigment solution and determine the absorption spectra of the pigments by using a spectrophotometer. Measure the absorbance of each pigment solution over the range of wavelength from 400–700 nm. Plot the absorption spectra.

13. MEASUREMENT OF THE RATE OF RESPIRATION USING GANONG'S RESPIROMETER

Background

In germinating seeds, carbohydrate is usually the main respiratory substrate and the respiratory quotient is usually 1 (as the amount of CO_2 released and O_2 absorbed is equal). When respiration takes place in a closed chamber of the Ganong's respirometer's bulb, the oxygen is continuously depleted from the atmosphere of the bulb and the CO_2 is added. When KOH is suspended in the bulb, KOH absorbs CO_2, thereby creating a partial vacuum in the bulb. To fill the vacuum, the water level rises up.

Principle

Respiration is a catabolic process, which involves the breakdown of complex substances into simple substances with a simultaneous release of energy. The process of respiration involves the utilization of oxygen with the release of carbon dioxide. Germinating seeds have a high rate of respiration, which can be calculated by using Ganong's respirometer. Generally, in plants carbohydrates are the most important substrates. However, some seeds also contain stored fats and proteins. The general equation for respiration is:

$$C_6 H_{12} O_6 + 6H_2O+6\,O_2 \rightarrow 6CO_2+ 12H_2O+ \text{Energy}$$

Materials

Pre–soaked seeds, water, thread, 1% potassium hydrogen oxide solution

Glassware and Equipments

Ganong's respirometer (Fig. 6), rubber tube, wooden stand, a small tube

Procedure

Take 25 pre–soaked gram seeds in the bulb of Ganong's respirometer. Before closing the mouth of the bulb with stopper, add 1%

KOH solution in the small tube present in the bulb. Fill the water in the tube and note down the initial level of water. Observe the change in water level at the interval of one hour. The rise in water level is direct measurement of the volume of O_2 consumed during respiration by the germinating seeds. Rate of respiration can be expressed as:

Increase in water level = Volume of O_2 consumed / Time (h) x Number of seeds

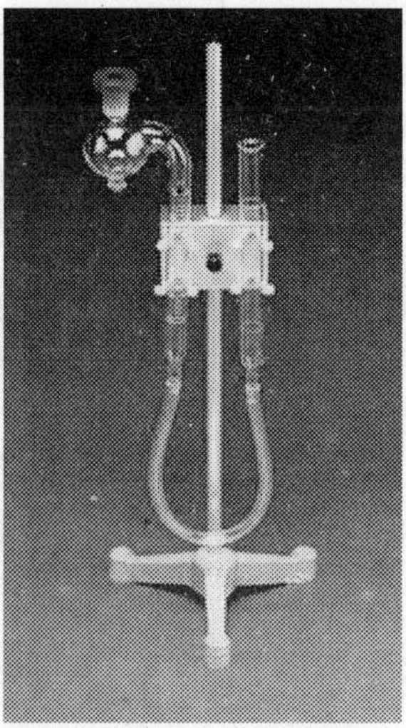

Fig. 6. Ganong's respirometer

Precautions

- Pre–soaked seeds should be used.
- The bulb should be airtight otherwise there will be no rise in the water level.
- Water level rise should not lead the water inside the bulb.
- KOH solution should be accurate.

14. DETERMINATION OF RESPIRATORY QUOTIENT (RQ) OF DIFFERENT TYPES OF RESPIRATORY SUBSTRATES

Materials

10% KOH solution, saline water or mercury, for starch - germinating seeds of wheat and maize, for fats- germinating seeds of mustard with testa removed, for organic acid - succulent plants such as *Opuntia* and *Euphorbia,* and for protein - germinating seeds of pulses with testa removed

Glassware and Equipments

Ganong's respirometer, beakers, stand

Procedure

Set the Ganong's respirometer on a stand. Fill saline water from the side of levelling tube (Saline water does not absorb any gas). Place moist seeds of one type of respiratory substrate in the bulb of respirometer. Make the level in the graduated tube parallel to the levelling tube. Adjust the level with the help of a levelling tube. Make the apparatus airtight by using the stopcock of the respiratory bulb. Record the initial level of saline water. Keep the experiment for a fixed time and record the change in level. Now, introduce a little KOH in saline solution. The KOH diffuses and absorbs carbon dioxide present inside the bulb, due to which level changes again. Record these changes. Record the whole process for all the different types of respiratory substrates and record the observations separately.

Observations and calculations

In case of starchy seeds, the first reading shows no change in level of saline solution in graduated tube. It indicates that the volume of O_2 absorbed is equal to the volume of CO_2 released during respiration. After the addition of KOH, the second reading however, shows a rise in the level due to absorption of CO_2 by KOH present in respiratory chamber.

In case of proteinaceous seeds and fatty acids, the first reading shows some rise in the level indicating that the lesser volume of CO_2 has been produced as compared to the volume of O_2 absorbed. The second reading, after introducing KOH, shows a further rise in the level due to absorption of CO_2 by KOH.

In case of organic acids, the first reading shows a fall in the level of solution indicating that the volume of CO_2 produced during respiration is more than the volume of O_2 absorbed. The second reading, after introducing KOH, shows a rise in the level above the original level. In succulent plants, the first reading shows a rise in the level of saline solution indicating that O_2 has been absorbed during respiration but little or no CO_2 has been released. The second reading, after the addition of KOH, shows no change in the level indicating that CO_2 has not been produced at all i.e., release of CO_2 is zero. It can be calculated as follows:

V_1 = Original level of saline solution

V_2 = First reading taken before the addition of KOH

V_3 = Second reading taken after addition of KOH

RQ of carbohydrate = $V_3–V_1/V_3–V_1$ = 1 or unity

RQ of proteins and fats = $V_3–V_2/V_3–V_1$ = Less than unity

RQ of organic acid = $V_3–V_2/V_3–V_1$ = More than unity

RQ of succulent plants = $V_2–V_2/ V_2–V_1$ = 0

Precautions

- Pre–soaked seeds should be used.
- The bulb should be airtight otherwise water level will not increase.
- Water level should not be too high to move inside the bulb.

15. DETERMINATION OF SEED VIABILITY AND VIGOUR TEST

Background

A viable seed is one, which is capable of germinating under the proper circumstances. Such a viable seed may or may not be readily or immediately germinable. Dormant viable seeds may require lengthy specific treatments before they become immediately germinable. Tetrazolium or TZ test is very useful for rapidly obtaining an indication of germination potential and viability of samples.

15.1. Tetrazolium Test

Principles

A colourless solution of 2,3,5-triphenytetrazolium chloride reacts with hydrogen released by the reduction process in living cells due to the action of enzyme dehydrogenase. It produces a red, stable and non-diffusible substance triphenylformazone to distinguish stained living cells from the colourless dead ones.

Materials

Seeds, water

Reagents

1. Phosphate buffer solution (pH 7.0) - Solution 1: Dissolve 9.087 g of KH_2PO_4 in 1,000 ml of water. Solution 2: Dissolve 11.876 g of $Na_2HPO4.2H_2O$ in 1,000 ml of water. Take 400 ml of Solution 1 and 600 ml of Solution 2, and mix them together.
2. Tetrazolium (TZ) solution (1%) - Dissolve 10 g of TZ salt in a litre of buffer solution (prepare fresh and keep in brown bottle).

Glassware and Equipments

Beaker, petri dish, foreceps

Procedure

Take the seeds and soak them in water overnight at room temperature. Next day cut the seeds longitudinally to expose the embryo (in case of monocots). In case of dicots, remove the seed coat to facilitate the quick penetration of tetrazolium. Keep the seeds in petri dish and soak in 1% tetrazolium solution and keep in dark at 30°C for 3–4 h for colour development. When the colour is developed, drain out the TZ solution, rinse the seeds 2–3 times with water, and evaluate the staining pattern. During evaluation seed should be immersed in water.

Observations

Red stained area indicates living tissues, while unstained area represents dead tissues. If entire embryo stains bright red in colour, it indicates that seed is viable. The intensity of red colour in the stained seed during TZ test depends upon the amount of formazan formed, which in turn is regulated by the dehydrogenase activity of the living cells. Therefore, in vigorous seeds, the intensity of the colour is expected to be more. Group the seeds into three categories - (1) completely–stained seeds, (2) completely–unstained seeds, and (3) partially–stained seeds, and calculate per cent seed viability.

Precautions

- The tetrazolium solution should always be kept in dark.
- As far as possible do the test in dark.

15.2. Germination Test

Background

Seed viability is an important aspect. The viability of the seed lot is a measure of how many seeds are alive and could develop into plants, which will reproduce themselves, given the appropriate conditions. The most accurate test of viability is the germination test using petri dish.

Materials

Filter paper, water

Glassware and Equipments

Petri dishes, forceps, incubator

Procedure

Take the filter paper and cut it according to the size of petri dish. Place a layer of paper in each dish. If the paper is too thin, then use a double layer. Label each petri dish with the sample number, replication and date. Moisten the paper with water. Take 200 seeds randomly for the test. Divide these seeds into at least two replicates. Arrange the seeds in a regular equidistant pattern on the surface of the paper. Cover it and keep it in the incubator for the definite time depending upon the crop.

Observations

After the incubation period, count the number of fully–formed seedlings. Normal seedlings are those, which have well–developed root system, shoot system, and are healthy (not stunted). Estimate the germination in percentage as:

Germination (%)=[Germinated seeds / Total number of seeds] x 100

16. EFFECT OF TEMPERATURE ON SEED GERMINATION IN CROP PLANTS

Materials

One or more types of crop seeds such as wheat, sweet corn, tomato, various beans, radish, peas, paper towels, water, resealable plastic bags

Glassware and Equipments

Tray, thermometers

Procedure

Determine the optimum temperature range for seed germination for each crop and take 60 seeds of each crop. Get three equal–sized sheets of paper towels. Prepare a seed germination chamber by folding the paper towel in half. With a pencil (not pen), label the outside of each towel as warm, cool, and room temperature. Place two rows of test seeds (total 20 seeds) inside the paper towel and fold it properly. Keep the labelled paper towels in a tray containing water of different for creating warm cool and room temperature environment. Record the number of seeds germinated each day as well as take the temperature of each towel. Arrange the recorded data into tables showing dates, temperature, number of germinated seeds, and percentage of germinated seeds. Plot a line graph of number of seeds germinated against the number of days for each temperature.

Precautions

- Be sure to check the towel moisture each day.
- Maintain a moist but not overly moist condition.
- The seeds should be selected randomly.

17. EFFECT OF LIGHT AND KINETIN ON SEED GERMINATION

Background

Seeds of many plants do not germinate immediately after ripening even under favourable conditions of germination such as appropriate moisture, temperature and oxygen. Such dormancy results from different conditions prevailing internally in the seeds and can be overcome by providing special treatments. One such treatment, which helps overcome dormancy in many seeds, is light treatment. The germination of many seeds is strongly influenced by light. Seed germination is promoted by light in some plants such as lettuce (*Lactuca sativa*) and tobacco (*Nicotiana tabacum*), while in some plants seed germination is inhibited by light, e.g., cucumber (*Cucurbits pepo*). Germination of some seeds is unaffected by light. In many cases of seed germination, e.g., in lettuce, that require light treatment for successful germination, light can be replaced by the phytohormone kinetin, leading to the speculation that the light effect on seed germination involve this hormone. However, light cannot always be replaced by kinetin in all kind of light–mediated processes. Therefore, light effect cannot be explained by hormone action alone.

Materials

Seeds of *Lactuca sativa* var. Grand rapid (incubate dry seeds at 35–37°C for 3 days before experiment), Summer squash (*Cucurbita pepo*), kinetin solutions (0, 1, 2, 5 and 10 ppm), filter paper

Glassware and Equipments

Petri dishes, light source

Procedure

Sow two benches of 50 seeds each of *Lactuca sativa* and *Cucurbita pepo* in petri dishes lined with filter paper moistened with 10 ml of test solution. Place one set of seeds under diffused light in the laboratory and place the other set in the dark. Similarly, take 50 *Lactuca* seeds in each petri dish containing different concentration of kinetin solution (0, 1, 2, 5, 10 ppm) and place them in dark. Count the number of seeds germinated after 4–5 days.

Results

Percentage of seed germination will be higher in light in case of *Lactuca sativa,* whereas in *Cucurbita pepo,* the seed germination will be higher in dark. Observe if kinetin can replace light requirement for promoting seed germination in *Lactuca.*

18. METHODS OF BREAKING SEED DORMANCY

Background

Inability of a viable seed to respond to the favourable environmental conditions for germination is known as seed dormancy. In this condition, seed is not able to germinate. Dormancy is generally of three types - hard–seededness, physiological dormancy and presence of inhibitory substances. It is possible to break these dormancies by applying physical and chemical method.

18.1. Methods of Breaking Hard–seededness

Hard–seededness is the result of the impermeability of the seed coat to water or gases or physical resistance to embryo expansion. This type of dormancy can be broken as:

1. **Seed soaking:** Dormant seeds are soaked in distilled water for 24–48 h and germination test is performed just after soaking, for example, Rice.
2. **Mechanical scarification**: Seeds of some crops, e.g., faba bean and rice are scarified mechanically to break the hard–seededness by the following method:
   a. **Piercing**: Seed coat is pierced with the help of a sharp needle without any injury to other parts of the seeds.
   b. **Chipping**: Seed coat is chipped with a sharp razor without any injury to other parts of the seed.
   c. **Filing**: Seed coat is scratched at the suitable site with a file or sand paper for entry of water.
3. **Acid scarification:** Seeds are soaked for a prescribed period in concentrated sulphuric or nitric acid to make the seed coat permeable to water, gases or physical resistance. After treatment, seeds are thoroughly washed under running tap water for several times and then tested for germination, e.g., cotton and rice.

18.2. Methods of Breaking Physiological Dormancy

Failure of viable mature embryo to germinate even when it is isolated from the seed coat is considered as physiological dormancy or embryo dormancy. This phenomenon is common in the woody members of family *Rosaceae*. It can be overcome by following methods:

1. **Dry storage**: Seeds of the crops having dormancy for short period of time are safely stored up to that period to break the dormancy, e.g., sunflower.

2 **Stratification**: Exposing imbibed seeds to cool or warm temperature prior to germination in order to break the dormancy is known as stratification.

3 **Pre–chilling treatment**: Seeds are arranged on a paper towel for germination test. These moistened paper towel containing seeds are kept at a low temperature (5–10°C) up to 7 days before germination test, e.g., lentil, sunflower, onion, oat, cauliflower, cabbage, barley, linseed, Lucerne, radish, sorghum, wheat, faba bean, lettuce, opium, spinach, onion, cole crop and turnip.

4. **Pre–heating treatment**: Seeds are arranged on a paper towel for germination test. These rolled towels are heated by free air circulation with maximum 30–35°C temperature for 7 days before germination, e.g., groundnut (40°C), wheat, oat barley, sunflower and berseem (30–35°C), and rice (50°C).

5. **Light** : Light with 750–1250 lux intensity from cool white lamp is illuminated for 8 h in every 24 h during the high temperature period to break the dormancy of grass seeds with alternate temperature, e.g., kalanchoe.

6. **Potassium nitrate**: A solution of potassium nitrate (0.2%) is prepared by dissolving 2 g KNO_3 in 1*l* of water. The substratum selected for germination test is moistened with this solution instead of water. Seeds are then kept for germination in the germinator, e.g., *Brassica*, tomato, cole crop and chilli.

7. **Gibberellic acid (GA_3)**: GA_3 solution of 0.05% is prepared by dissolving 500 mg GA_3 in 1 *l* water. Seeds are arranged on a substratum moistened with this solution and placed in the germinator, e.g., oat, barley and wheat.

18.3. Methods for Removal of Inhibitory Substances

Some crops develop inhibitory substances on the surface of the seeds at the time of physiological maturity in the field. These inhibitory substances have the ability to check the germination of viable seeds. Removal of these substances by following methods from the seed results in normal germination:

1. **Pre–washing**: Inhibitory substances from the surface of the seed coat are removed by washing the seed in running water at 25°C. Seeds are then dried at room temperature prior to germination test, e.g., Beta, *Xanthium* and Fraxinus.

2. **Removal of structure around the seed**: Inhibitory structures such as involucres of bristle, lemma and palea and hard shell are removed with the help of forceps prior to germination.

19. MEASUREMENT OF LEAF AREA

Background

Leaf area is an important measurement in bioproductivity studies, as it bears important and direct relationship to photosynthesis and transpiration. Leaf area is also important in determining the percentage of solar radiation intercepted by an individual plant or crop canopy. Photosynthesis depends on individual leaves. However, productivity of vegetation depends not only on individual leaves but on canopy structure. The photosynthetic productivity of a canopy depends on both the number of leaves present and the configuration of the leaves making up the canopy. This leaf area directly influences plant growth and the final yield. The 'Leaf Area Index' (LAI) or leaf area per unit land area is computed as:

$$LAI = La/P$$

Here, La is the functional leaf area (green leaves) of the crop canopy standing on ground area P. As both La and P are normally measured in m^{-2}, LAI is dimensionless. In field studies, the leaf area index forms the basis of the relationship between absorbed radiant flux and plant growth.

The leaf area can be measured by any of the following methods:

- Graph paper method
- Paperweight method
- Leaf area meter

19.1. Graph Paper Method

Materials

Sample leaves, graph paper, pencil

Procedure

Trace the leaves on the graph paper and count the squares covered to give you an estimate of the surface area for each leaf. Repeat this for each leaf on a plant and for each plant in the experiment.

Observations

Count the squares of estimated leaves and calculate the leaf area per leaf or plant. The estimated leaf area of sample is expressed as cm^2.

Precautions

- The leaf should be fresh and without folding.
- The leaf should not be thorny.

19.2. Paperweight Method

Materials

Sample leaves, paper, pencil, scissor

Procedure

Trace out each leaf on paper. Cut out the leaf tracings and weigh them. Weigh the cut outs and divide the total weight by the number of leaves to give the average leaf area for each plant. Repeat this for each of the plants in your experiment. The leaf area of the given sample or plant can be expressed in cm^2.

Precautions

- Make sure to use the same type of paper every time.
- The paper should not be wet.

19.3. Leaf Area Meter

This is the most advanced method for the measurement of leaf area. It is designed for laboratory measurement of large and small leaves. Direct measurements of leaf area within electronic Area meter generally provides the lowest coefficient of variance and requires the fewest plants to be measured in order to detect comparable differences among crop canopies.

Procedure

Samples are placed between the guides on lower transparent belt and then passed through the LI-3100. As the samples travel under the 15 watt fluorescent light source, the accumulating area in cm^2 is shown on LED display. The leaf image is reflected by a system of three mirrors to a solid scanning camera within the rear housing. The mirrors are mounted on the instrument base plate. Two are visible under the transparent belts. The third is within rear housing.

Calibration

Before using the instrument, it needs a proper calibration. For this, measure a paper of known area. Cut the paper into many small discs and take a reading. Adjust the "CAL" screw on the front panel until the area reading for small discs equals the whole paper reading. The difference in the area is an estimate of the inherent error due to a large ratio of edge to surface area in the case of the cut paper. Finally, take leaf area measurements.

Precautions

- Allow instrument to warm up for at least 30 min prior to calibration.
- Check calibration frequently.
- Loose belt or stretched belt causes spurious counts. Tighten or replace the belt.
- Fluorescent tube may not be illuminated. Press 'LAMP START' switch.
- Turn screw counterclockwise until display number remains constant.

20. MEASUREMENT OF GROWTH ANALYSIS

Background

Growth can be defined as irreversible increase in size (in terms of cell division, enlargement and differentiation or increase in dry weight). The pattern of growth over a generation (growth cycle) is typically characterized by a growth function referred to as the sigmoid curve. The sigmoid accumulation pattern typifies all organisms, organs, tissues and even cell constituents. The pattern of sigmoid growth is explained in four phases - (1) Exponential growth phase, (2) Linear phase, (3) Phase of declining growth rate, and (4) Phase of steady state.

Growth analysis can be defined as an useful approach to analyse production by plant or plant parts. Growth analysis is the first step in the analysis of production and is a link between production and growth.

The study of growth analysis provides information on the growth of plant or plant community in a precise way with one on the nature of plant and environment interaction in a particular habitat. Two types of primary data are usually required for the growth analysis. These are:

(1) Plant dry weight

(2) The size of assimilatory system i.e. leaf area.

The primary data of growth analysis can be collected from individual plants or derived from the whole canopies. With the help of primary data growth is described quantitatively. The quatitative description of growth is based upon several terminology which we call as growth parameters component of growth analysis.

20.1. Relative Growth Rate (RGR)

This is the basic component of growth analysis, which arose from the work of Blackman (1919). RGR is defined at any instant of time (t) as the increase in dry weight per unit dry material present. This can be determined by measuring plant dry weight periodically during growth and is commonly represented as g g^{-1} day^{-1} or g g^{-1} week.

$$RGR = \ln W_2 - \ln W_1 / t_2 - t_1$$

Or

$$RGR = (\log 10\ W_2 - \log 10 W_1) \times 2.303 / t_2 - t_1$$

Where, W_1 and W_2 are the plant dry weight at times t_1 and t_2 respectively.

Drawback: RGR does not give any knowledge about the size of assimilatory surface and the amount of photosynthate that is going into the plant material. So in succeeding years two other concepts were developed for use in growth analysis. These were NAR and LAR.

$$RGR = NAR \times LAR$$

20.2. Net Assimilation Rate (NAR)

NAR is defined as the increase in plant dry weight per unit of assimilatory surface per unit time. NAR can be determined by measuring plant dry weight and leaf area periodically during growth and is commonly reported as g cm^{-2} $week^{-1}$

$$NAR = (W_2 - W_1)(\ln A_2 - \ln A_1) / (t_2 - t_1)(A_2 - A_1)$$

Or

$$NAR = (W_2 - W_1)\ 2.303\ (\text{Log } 10\ A_2 - \text{Log } 10 A_1) / (t_2 - t_1)(A_2 - A_1)$$

Where, W_1 and W_2 are the plant dry weight at times t_1 and t_2 respectively. Similarly, A_1 and A_2 are the leaf area at times t_1 and t_2.

20.3. Leaf Area Ratio (LAR)

LAR of a plant at any instant of time (t) is the ratio of total assimilatory surface to whole plant dry weight. This can be determined by measuring total leaf area and whole plant dry weight periodically and is commonly reported as $cm^2\ g^{-1}$.

$$LAR = (A_2 - A_1)(\ln W_2 - \ln W_1) / (W_2 - W_1)(\ln A_2 - \ln A_1)$$

Or

$$LAR = \frac{(A_2 - A_1)\ 2.303\ (\log 10\ W_2 - \log 10\ W_1) / (W_2 - W_1)}{2.303\ (\log 10\ A_2 - \log 10\ A_1)}$$

Where, W_1 and W_2 are plant dry weight at times t_1 and t_2. Similarly, A_1 and A_2 are leaf area at times t_1 and t_2.

Further, LAR is the product of leaf weight ratio and specific leaf area.

$$LAR = LWR \times SLA$$

20.4. Leaf Weight Ratio (LWR)

LWR is the ratio of dry weight of leaves to whole plant dry weight and is commonly reported as ($g\ g^{-1}$).

$$LWR = WL / W$$

Where, WL = dry weight of leaves; W = plant dry weight

20.5. Specific Leaf Area (SLA)

SLA is a measure of the leafiness of the plant on a dry weight basis and is commonly reported as ($cm^2\ g^{-1}$). It is the ratio of leaf area to leaf dry weight.

$$SLA = \text{Leaf area} / \text{Leaf dry weight}$$

20.6. Absolute Growth Rate (AGR)

AGR is the function of amount of growing material present and is influenced by the environment. It gives absolute values of biomass between two intervals. It is mainly used for a single plant or single plant organ, e.g., leaf growth, plant height.

20.7. Crop Growth Rate (CGR)

It is defined as the rate of primary production per unit land area per unit time (Watson, 1952).

$$CGR = W_2 - W_1 / (t_2 - t_1) \times 1 / L$$

Where, W_1 and W_2 are plant dry weight at time t_1 and t_2 and L is the land area.

$$CGR = LAI \times NAR$$

20.8. Leaf Area Index (LAI)

LAI may be defined as the area of leaves (one side only) divided by ground area over which it is growing. Since leaf area and land area both have the same units, LAI is dimensionless.

$$LAI = A / L$$

Where,

A = Total leaf area

L = Ground area

20.9. Leaf Area Duration (LAD)

This is the measure of persistence of the assimilatory surface. As this is the product of LAI and time, the unit of LAD is days.

$$LAD = (A_1 + A_2)(t_2 - t_1) / 2$$

Where, A_1 and A_2 are the leaf area index at times t_1 and t_2, respectively. An approximate estimate of total yield can now be defined as the product of LAD x NAR.

20.10. Harvest Index

It is the ratio of economic yield to the total biological yield expressed in percentage. It represents the efficiency of photosynthate translocation to the economic parts. Economic yield is the amount of productivity, which is partioned into the useful or harvested portion of the crop, e.g., grains of cereals, the trunk of timber tree or the shoots of herbage crops.

$$\text{Harvest Index} = (\text{Economic yield} / \text{Biological yield}) \times 100$$

21. PLANT NUTRITION AND MINERAL DEFICIENCY

Background

In addition to carbon, hydrogen and oxygen, plants also require other elements, which they absorb from the soil in the form of inorganic ions. Chemical analysis of plants reveals the presence of about 60 mineral elements, out of which only 17 elements are considered to be essential for plant growth and development. Some elements such as nitrogen, phosphorous, potassium, calcium, magnesium and sulphur are required relatively in large concentrations. Such elements are called as macroelements, whereas elements such as iron, copper, zinc, molybdenum, boron, manganese chlorine and nickel are required in traces and are called minor or microelements. Deficiency of a particular element leads to the development of characteristic symptoms in plants. In this experiment, we will study the deficiency symptoms associated with the deficiency of a particular element. If a key to mineral deficiency symptoms is prepared, it can enable the agriculturist to diagnose the symptoms and correct them accordingly. A key to nutrient deficiency symptoms can be prepared by growing plants in sand culture and solution culture by deleting one of the essential elements from the complete nutrient solution.

Principle

The principle of nutrient solution is based on the fact that plants are nourished with nutrient solution containing all the essential elements except the one in question. The growth of the plant in such solution is then compared with that of plants provided with complete nutrient solution. The conclusion relating to omitted nutrient is based on the difference in growth of the two sets of plants and disorder developed.

Materials

9–12 days old seedlings of tomato, bean, beakers, measuring cylinder, glass bottles of about 1 *l* capacity, metal lids or flat cork, stock solutions of macro and micronutriens.

Procedure

Preparation of nutrient solution

Macronutrients : 1 M stock solution of inorganic salts: KH_2PO_4, KNO_3. Ca $(NO_3)_2$, $MgSO_4$, KCl, $CaCl_2$, Na_2SO_4, $NaNO_3$, $MgCl_2$

Micronutrients : Weigh the following salts in the quantity mentioned against each. Dissolve them in little water and make the volume exactly 1 *l*. Use 1 ml of supplementary solution per litre of nutrient solution.

H_3BO_3 - 2.86 g/l, $MnCl_2.4H2O$ - 1.81 g/l, $ZnSO_4.7H_2O$ - 0.22 g/l, $CuSO_4.5H_2O$ - 0.08 g/l, H_2MoO_4 - 0.02g/l

Iron solution - Weigh 36.5g of Fe–EDTA, dissolve it in little water and make the volume exactly 1 *l*. Use 1 ml iron solution per litre of nutrient solution.

Take glass bottles of uniform size having metal lid or cork. The outer side of the bottles should be covered with black paper to prevent inhibitory effect on root growth. Now, add the volume of each stock solution as indicated in Table 2. Stir the solution with a glass rod upon addition of each component to prevent precipitation. When required amount of stock solutions are added to each bottle, label as complete, nitrogen lacking, phosphorous lacking and so on. Cover the top of the glass bottle with perforated circular metal caps, corks or cork boards to provide support to the seedlings as well as to provide aeration to the nutrient solution.

Now keep one or two small tomato seedlings in each jar with the support of cover and keep the complete set up in the greenhouse. The nutrient solution should be completely aerated through bubbling with the help of air pump.

Table 2. Composition of nutrient solution

Stock solution	**1** +	**2** -N	**3** -K	**4** -P	**5** -Ca	**6** -Mg	**7** -S	**8** -Fe	**9** -Mn
1M $CaNO_3$	5 ml	-ml	5 ml	5 ml	-ml	5 ml	5 ml	5 ml	5 ml
1M KNO_3	5	-	-	5	5	5	5	5	5
1M $MgSO_4$	2	2	2	2	2	-	-	2	2
1M KH_2PO_4	1	1	-	-	1	1	1	1	1
Fe-EDTA	1	1	1	1	1	1	1	-	-
MN	1	1	1	1	1	1	1	1	1
1M $NaNO_3$	-	-	5	-	10	-	-	-	-
1M $MgCl_2$	-	-	-	-	-	-	2	-	-
1M Na_2SO_4	-	-	-	-	-	2	-	-	-
1M NaH_2PO_4	-	-	1	-	-	-	-	-	-
1M $CaCl_2$	-	5	-	-	-	-	-	-	-
1M KCl	-	5	-	1	-	-	-	-	-

The number indicated above represent the volume of stock solution to be used per litre of nutrient solution.

+ = Complete medium with no deficiency

MN = Micronutrient

Observations

Carefully observe and note the deficiency symptoms. Take the photograph of plants showing deficiency symptoms. Also record the final plant height and plant mass in each treatment with the key to nutrient deficiency symptoms for confirmation.

Deficiency symptoms

Nitrogen - The most easily observed symptom of nitrogen deficiency is yellowing (chlorosis) of leaves due to loss in chlorophyll. The symptoms first appear on older leaves and then in the upper ones i.e., more actively growing leaves because of high mobility of nitrogen in plants.

Phosphorus - Many of the symptoms of phosphorus deficiency can be confused with nitrogen deficiency. Similar to nitrogen deficiency phosphorus deficiency may cause premature leaf fall and purple or red anthocyanin pigmentation. It develops dead necrotic spots on leaf, stem and fruits. The leaf may have a characteristic dark to purple green colouration. Because of high mobility of phosphorus in the plants, the older leaves first exhibit deficiency symptoms.

Calcium - Meristematic region at stem, leaf and root tips are greatly affected and eventually die, thus terminating growth in these organs. Chlorosis occurs along the margin of younger leaves and this area usually becomes necrotic. Malformation or distortion of the younger leaves is also due to the calcium deficiency in plants, a hooking of the leaf tip being the most easily detected symptoms. Deficiency first appears in the younger leaves and the growing apices, probably as a consequence of immobility of calcium in plants.

Magnesium - It causes the interveinal chlorosis of leaves. Yellowing is first apparent on basal leaves, and as the deficiency becomes more acute, eventually it reaches the younger leaves. It is mobile in nature. In severe conditions, it forms necrotic spots.

Potassium - A mottled chlorosis first occurs, which is followed by the development of necrotic areas at the tip and margin of the leaf. Generally, these symptoms appear first on the more mature leaves due to the mobility of the potassium. There is a tendency for the leaf tip to curl downward and as in case of the French bean and potato, marginal region may roll inward.

Sulphur - Similar to nitrogen deficiency, sulphur deficient plants also show general chlorosis followed by the production of anthocyanin pigmentation in some species. Sulphur deficient plants show chlorosis first on the younger leaves. Under severe conditions, however, all of the leaves may undergo some loss of green colour.

Iron - The most easily observed symptom of iron deficiency in the plants is extensive chlorosis in the leaves. The younger leaves are most affected, the more mature leaves sometime show no chlorosis at all. Primarily, this is because of the relative immobility of iron in plants. Thus, the younger leaves cannot withdraw iron from the older leaves. One feature of iron–induced chlorosis is its characteristic interveinal nature.

Manganese - Manganese deficiency shows chlorotic and necrotic spots in the interveinal areas of the leaf. These symptoms may first appear on the younger leaves of some species, while in other species they may appear first on the older leaves.

Copper - Copper deficiency causes a necrosis of the tip of younger leaves that proceeds along the margin of the leaf and gives a withered appearance. Under more severe conditions, the leaves may be lost, and the whole plant may appear wilted.

Zinc - The first sign of zinc deficiency is an interveinal chlorosis of the older leaves, starting at the tip and margins, followed by necrotic spots. Smaller leaves and short internodes, resulting in stunted growth, are more severe zinc–deficiency symptoms. Plants are generally smaller in size, distorted in shape and appearance, and may be clustered on short branches known as rosettes.

Boron - The first visible symptom of boron deficiency, which occurs only a few hours after boron withdrawal, is the death of the shoot tip because of its requirement for DNA synthesis. The tip of these shoot also die. The flowers do not form and root growth is stunted.

Molybdenum - Visible symptoms of molybdenum deficiency may start with chlorotic interveinal mottling of the lower leaves, followed by marginal chlorosis and unfolding of the leaves. Under more severe conditions, mottled area may become necrotic and may cause the leaf to wilt. Flower formation is inhibited and if flowers do form, they abscise before setting fruits.

Precautions

- Replace the used solution with freshly prepared solution every alternate day or at weekly intervals as desired.

- Nitrogen should be applied both as ammonium nitrogen and nitrate nitrogen.

PART B - BIOCHEMISTRY

1 CARBOHYDRATES

Most substances of this class have empirical formula $C_{n\phi}$ $(H_2O)_n$ suggesting that they are the hydrates of carbon. Most of the carbohydrates confirm to this formula and some of them do not. There are some compounds, which fulfil above formula but they are not sugars. So, chemically carbohydrates are polyhydroxy aldehydes or ketones or all those substances, which on hydrolysis give such compounds.

Classification of carbohydrates

They are of three types:

1. Monosaccharides
2. Oligosaccharides
3. Polysaccharides

1. Monosaccharides: The word saccharide is derived from the Greek word 'sakkharon', meaning sugar. They are simple sugars, which cannot be further split into simpler sugars by hydrolysis. They consist of a single polyhydroxy aldehyde or ketone unit. The most abundant monosaccharide in nature is the six carbon sugar, D–glucose. They are colourless, crystalline solids that are freely soluble in water but insoluble in non–polar solvents. Most of them are sweet in taste.

Based on the position of carbonyl group, they are of two types. If the carbonyl group is at an end of the carbon chain, the monosaccharide is aldehyde and is called an aldose. If the carbonyl group is at any other position, the monosaccharide is ketone and is called ketose. Depending upon the number of carbon atoms, they can be triose (3C), tetrose (4C), pentose (5C), hexose (6C) and heptose (7C).

2. Oligosaccharides: They consist of short chain (2–10) of monosaccharide units joined together by characteristic glycosidic linkage. They can be di, tri, tetra saccharides. The most abundant are the disaccharides ($C_{12}H_{22}O_{11}$) with two monosaccharide units, e.g., sucrose (cane sugar)-consists of D–glucose and D–Fructose; Maltose- consists of two units of D–glucose; Lactose (milk sugar) - consists of one unit of D–

glucose and D–galactose each. Most oligosaccharides having three or more monosaccharide units do not occur in free form but exist as in bound form with lipids or proteins.

3. Polysaccharides: They consist of long chains of monosaccharide, may be hundreds or thousands. They may be of linear type such as cellulose or may be branched such as glycogen or starch.

a. **Starch:** It is the storage form of glucose in plants. Starch consists of amylose, a linear polymer of glucose, joined by α-1,4 linkage. Amylopectin is a branched polymer where α-1,4 linkage in linear chain and α-1,6 linkages at branch points are present. In starch, branching is less as compared to glycogen, occurring about once in every thirty units. It is hydrolyzed by amylase and gives rise to maltose units, maltotriose (has three glucose units) and α–dextrin (made of several glucose units joined by α-1,4 linkage and α-1,6 linkage). Maltose and maltotriose are hydrolyzed to glucose by maltase. α–dextrin is hydrolyzed to glucose by α–dextrinase. Malt contains β–amylase enzyme that hydrolyses starch into maltose by sequential removal of disaccharide units from non–reducing ends. Malt derived from barley or other grains is used to make beer.

b. **Glycogen:** It is the storage form in animal cells. Its structure is similar to starch but branches are more frequent as compared to starch and occur once in every ten units.

c. **Cellulose:** It is an unbranched polymer of glucose joined by β-1,4 linkage. It is the most abundant organic compound in the biosphere.

Qualitative Analysis

1.1. Molisch's Test

Principle

This is a general test for identification of carbohydrates. Concentrated sulphuric acid dehydrates sugars and forms furfurals (in case of pentoses) and hydroxyl methyl furfurals (in case of hexoses), which then combine with α–napthol to form coloured compounds.

Glassware and Equipments

Test tubes, test tube stand, pipettes, reagent bottles, dropper, volumetric flasks, electric balance

Reagents

1. Molisch Reagent: 5% solution of α–napthol in alcohol (Dissolve 5 g of α–napthol in ethanol and make the volume to 100 ml.)
2. Concentrated H_2SO_4
3. 1% sugar solution (glucose, fructose, sucrose, lactose and starch): Dissolve 1 g of each sugar in distilled water and make its volume to 100 ml in each case separately.

Procedure

Take 2 ml of dilute sugar solution, add 2 drops of Molisch reagent and mix the contents. Add 2 ml of concentrated sulphuric acid carefully by the side of tube in slanting position. Formation of reddish violet (purple) ring at the junction of two liquids shows presence of sugars.

Note

- Concentrated sugar solution will give red colour instead of violet due to charring of sugars by the acid. So, in that case dilute sugar solution should be tested.
- Trioses and tetroses do not give this test.

1.2. Iodine Test

Principle

Iodine forms a colour adsorption complex with polysaccharides such as blue with starch, red with dextrin, wine red with glycogen. These are not definite compounds. The colour is due to the formation of a coordination complex between helically coiled polysaccharide chain and iodine is centrally located within the helix.

Glassware and Equipments

Test tubes, test tube stand, pipettes, measuring cylinder, beakers, dropper, volumetric flask, electric balance

Reagents

1. Iodine reagent: Dissolve 2 g of potassium iodide (KI) to make 100 ml solution and add sufficient iodine to get deep yellow colour.

Procedure

Take 2–3 ml of dilute solution of a polysaccharide. Add 1–2 drops of freshly prepared iodine solution. Formation of coloured complex shows test is positive for polysaccharides.

Note

- In case of starch, on heating, helical structure of amylose is disrupted and it loses I_2 binding capacity. As a result, blue colour disappears. On cooling, colour reappears because of regaining of helical structure.
- Same does not happen in dextrin and glycogen.
- Increase in temperature decreases the colour.
- The colour depends on the length of the unbranched or linear (α-1,4) linked chain available for the complex formation.

1.3. Reduction Test

Carbohydrates with free aldehyde or ketone groups have the ability to reduce solutions of various metallic ions.

1.3.1. Fehling test

Principle

Alkaline solutions of copper are reduced by sugars having a free aldehyde or ketone group. In turn, aldose sugars are oxidized to their corresponding aldonic acids, whereas keto sugars give rise to shorter-chain acids. This reaction results in the formation of yellow or red coloured precipitates of cuprous oxide giving positive test for reducing sugars.

Glassware and Equipments

Test tubes, test tube stand, pipettes, beakers, volumetric flasks, Bunsen burner or heater, tongs, electric balance

Reagents

1. Fehling reagent A: Dissolve 34.65 g of $CuSO_4.5H_2O$ in 100 ml of distilled water.
2. Fehling reagent B: Dissolve 25 g KOH and 34.6 g sodium potassium tartarate (Rochelle salt) in distilled water and make its volume to 100 ml.

3. Fehling solution C: Freshly mix equal volumes of Fehling reagent A and Fehling reagent B before use.

Procedure

Take 1 ml of sugar solution. Add 1 ml of Fehling solution C and mix thoroughly. Heat the contents slowly to boil. Colour changes from bluish to yellow and later red. The change in colour shows test is positive for reducing sugars.

Note

- Chloroform, uric acid and creatinine interfere in this test.
- Ammonium salts also interfere with Fehling solution, so solution should be made alkaline with sodium carbonate and boiled in order to decompose ammonium salts.

1.3.2. Benedict Test

Principle

Alkaline solutions of copper are reduced by sugars having a free aldehyde or ketone group. In turn, aldose sugars are oxidized to their corresponding aldonic acids, whereas keto sugars give rise to shorter–chain acids. This reaction results in the formation of yellow or red coloured cuprous oxide giving positive test for reducing sugars. Benedict's reagent is modified Fehling solution, which is a single reagent and highly stable.

Glassware and Equipments

Test tubes, test tube stand, pipettes, beakers, volumetric flasks, Bunsen burner, tongs, dropper, electric balance

Reagents

1. Solution A: Dissolve 173 g of sodium citrate and 100 g of anhydrous sodium carbonate in 500 ml of distilled water while heating. Filter and add water to make up 850 ml.
2. Solution B: Dissolve 17.3 g of $CuSO_4.5H_2O$ in 100 ml of distilled water while heating, cool and dilute to 150 ml.
3. Benedict's reagent: Slowly add solution B to solution A with constant stirring. Cool and transfer to 1 litre flask.

Procedure

Take 1 ml of sugar solution and add 2–5 ml of Benedict's solution. Mix the contents thoroughly. Heat the mixture to boil and note the change in colour. Formation of yellow to red coloured precipitate indicates test is positive.

Note

- This test is very sensitive and can detect sugars up to 0.1%.
- Chloroform, uric acid or creatinine does not interfere in this test as these interfere in Fehling solution test.

1.3.3. Nylander's Test (Bismuth Reduction Test)

Principle

Carbohydrates with free aldehyde or ketone groups reduce Nylander's reagent to cause black precipitates of metallic bismuth.

Glassware and Equipments

Test tubes, test tube stand, pipettes, measuring cylinder, beakers, volumetric flasks, water bath, dropper, electric balance

Reagents

1. 10% KOH: Dissolve 10 g KOH in distilled water and make its volume to 100 ml.
2. Nylandar's reagent: Dissolve 2 g bismuth subnitrate and 4 g of sodium potassium tartarate (Rochelle salt) in 10% KOH with heating and make the volume to 100 ml.

Procedure

Take 5 ml of sugar solution. Add a few drops of Nylander's reagent. Heat for 5 min in a boiling water bath. Formation of black precipitates of metallic bismuth confirms the presence of reducing sugar.

Note

- This test can detect sugars up to 0.08 % concentration.
- In this test uric acid does not interfere.

1.3.4. Core's Test

Principle

Carbohydrate with free aldehyde or ketone groups reduces copper sulphate to cuprous oxide on boiling and forms a yellow precipitates.

Glassware and Equipments

Test tubes, test tube stand, pipettes, beakers, volumetric flasks, Bunsen burner, tongs, dropper, electric balance

Reagents

1. 5% $CuSO_4$: Dissolve 5 g $CuSO_4$ in distilled water and make its volume to 100 ml.
2. Anhydrous Na_2CO_3
3. Glycerol

Procedure

Take 5 ml of sugar solution. Add large quantity of anhydrous Na_2CO_3. Mix and add 2–3 drops of glycerol and 2–3 drops of 5% $CuSO_4$ solution. Boil the contents. On boiling formation of yellowish precipitates confirms presence of reducing sugar.

1.3.5. Methylene Blue Test

Principle

Carbohydrates with free aldehyde or ketone groups can reduce methylene blue to leucomethylene blue with discharge of blue colour.

Glassware and Equipments

Test tubes, test tube stand, pipettes, beakers, volumetric flasks, Bunsen burner, tongs, dropper, electric balance

Reagents

1. 1% methylene blue: Dissolve 1 g methylene blue in distilled water and make its volume to 100 ml.
2. 5% NaOH: Dissolve 5 g NaOH in distilled water and make its volume to 100 ml.

Procedure

Add one drop of 1% methylene blue and a few drops of 5% NaOH to 2 ml of water. Boil carefully so that the colour does not disappear. Now add a few drops of 0.2% sugar solution and boil again. Colour is now discharged. Shake and note the reoxidation of leucomethylene blue by O_2 in the air to the coloured dye again.

1.4 Barfoed's Test

Principle

This test is used to distinguish monosaccharide from reducing disaccharide. It differs from Benedict's and Fehling test in a way that reduction of copper is brought about in acidic conditions. The acid makes this reagent a weak oxidizing reagent, therefore any monosaccharide will reduce cupric to cuprous, and Cu_2O will be separated. The reducing disaccharide reduces this reagent slowly as compared to monosaccharide.

Glassware and Equipments

Test tubes, test tube stand, pipettes, measuring cylinder, beakers, volumetric flasks, Bunsen burner, tongs, dropper, electric balance

Reagents

1. Barfoed's Reagent: Take 13.3 g of copper acetate and dissolve in 200 ml of water, and to it add 1 ml of glacial acetic acid.

Procedure

Take 1 ml of sugar solution, add 5 ml of Barfoed's reagent. Mix the contents and boil for 30 seconds. Formation of red precipitate indicates presence of monosaccharides.

Note

- If heating is done for a few minutes, disaccharides are hydrolyzed giving this test positive.
- Concentrations of monosaccharide and disaccharides should not be too different, as concentrated disaccharides will reduce this reagent quickly.

1.5. Seliwanoff's Test

Principle

This test is used to distinguish fructose from glucose. Concentrated acid causes dehydration of sugars to form hydroxymethyl furfural and furfural. Ketohexoses (fructose) forms hydroxymethyl furfural more readily than aldohexoses (glucose), which forms the basis of this test. Hydroxymethyl furfural reacts further with resorcinol to form brownish red coloured precipitate.

Glassware and Equipments

Test tubes, test tube stand, pipettes, measuring cylinder, beakers, volumetric flasks, Bunsen burner, tongs, dropper, electric balance

Reagents

1. Seliwanoff's reagent: Dissolve 0.5 g of resorcinol in 33 ml of concentrated HCl and dilute to 100 ml with water.

Procedure

Take 5 ml of Seliwanoff's reagent. Add 1 ml of sugar solution. Mix thoroughly and boil for 20 seconds. Red colour with or without precipitates shows test is positive.

Note

- Fructose, both in free and combined (sucrose) form, gives this test positive as the reaction is carried in strong acidic conditions. Sucrose on hydrolysis yield glucose and fructose and hence, gives this test positive.
- Glucose or maltose can give similar colour if boiling is prolonged.
- If glucose is absent, test is sensitive for 0.1% of fructose. If glucose is present, sensitivity decreases as large amount of glucose can give a similar colour.

1.6. Test for Pentoses

1.6.1. Aniline Test

Principle

Pentoses undergo dehydration much more readily than hexoses. As a result furfural is produced by heating pentoses in acid solution. The furfural can condense with phenolic compounds to give coloured products or with amino compounds to form Schiff's bases.

Glassware and Equipments

Test tubes, test tube stand, pipettes, beakers, volumetric flasks, Bunsen burner, tongs, dropper, electric balance

Reagents

1. Aniline acetate reagent: Take 35 ml glacial acetic acid and 50 ml aniline in a conical flask. Add 85 ml of distilled water and mix thoroughly. Use freshly prepared reagent.
2. Aniline acetate paper: Take a few filter paper strips and put a few drops of aniline acetate reagent over it.
3. Concentrated HCl

Procedure

Take 1 ml of sugar solution and add 1 ml of concentrated HCl. Keeping the test tube in slanting position, heat the contents of the test tube slowly to boil. Hold a filter paper dipped in aniline acetate reagent above the test tube so that vapours touch the filter paper. Formation of red colour indicates presence of pentose sugar.

1.6.2. Bial's Test

Principle

Pentoses undergo dehydration much more readily than hexoses. As a result, furfural is produced by heating pentoses in acid solution. The furfural can condense with phenolic compounds to give coloured products or with amino compounds to form Schiff's bases.

Glassware and Equipments

Test tubes, test tube stand, pipettes, measuring cylinder, beakers, volumetric flasks, Bunsen burner, tongs, dropper, electric balance

Reagent

1. 10% $FeCl_3$: Dissolve 10 g $FeCl_3$ in distilled water and make its volume to 100 ml.
2. Bial's reagent: Dissolve 300 mg of orcinol in 100 ml of concentrated HCl and add 5 drops of 10% $FeCl_3$.

Procedure

Take 0.5 ml of sugar solution. Add 5 ml of Bial's reagent. Boil the contents. Development of green colour shows test is positive for pentose.

Note

- Test is sensitive to 0.1 percent of pentose.
- Glucouronates on heating are converted to pentoses so they can give similar colour.

1.6.3. Tauber's Test

Principle

Pentoses undergo dehydration much more readily than hexoses. As a result furfural is produced by heating pentoses in acid solution. The furfural can condense with phenolic compounds to give coloured products or with amino compounds to form Schiff's bases.

Glassware and Equipments

Test tubes, test tube stand, pipettes, measuring cylinder, beakers, volumetric flasks, Bunsen burner, tongs, dropper, electric balance

Reagents

1. Benzidine

Procedure

Take 1 ml of Benzidine. Add 2 drops of sugar solution. Boil the contents and cool quickly. Formation of violet colour shows test is positive for pentose.

1.7. Hydrolysis Test for Sucrose

Principle

Sucrose, on hydrolysis by HCl, is converted into glucose and fructose. The presence of these two monosaccharides is detected by different tests.

Glassware and Equipments

Test tubes, test tube stand, pipettes, measuring cylinder, beakers, volumetric flasks, Bunsen burner, tongs, dropper, electric balance

Reagents

1. Dilute HCl
2. 2% Na_2CO_3 - Dissolve 2 g Na_2CO_3 in distilled water and make final volume to 100 ml.

Procedure

Take 5 ml of sucrose solution. Add 2 drops of dilute HCl and one drop of thymol blue. Acidic solution thus formed is pink in colour. Divide above solution in two parts. Boil one portion for about 1 minute and cool it under the tap. Keep second portion as it is. Neutralize both portions by adding 2% Na_2CO_3 solution drop wise. Formation of blue colour indicates solution is neutralized. Sucrose in the boiled portion gets hydrolyzed to glucose and fructose, which can be detected by Benedict's solution and Seliwanoff's reagent, respectively. Other unboiled portion does not reduce Benedict's solution.

1.8. Methylamine Test for Lactose

Principle

When the disaccharides lactose and maltose are hydrolyzed under alkaline conditions, they form enediols. The enediol reacts with methylamine hydrochloride to form red–coloured complex.

Glassware and Equipments

Test tubes, test tube stand, pipettes, measuring cylinder, beakers, volumetric flasks, waterbath, tongs, dropper, electric balance

Reagents

1. 0.2% methylamine solution: Dissolve 0.2 g methylamine hydrochloride in distilled water and make its volume to 100 ml.
2. 10% NaOH: Dissolve 10g NaOH in distilled water and make its volume to 100 ml.

Procedure

Take 5 ml of sugar solution. Add 1 ml of methylamine solution. Then add 0.2 ml of NaOH solution. Mix the contents. Cover the tubes with a glass bulb and heat at 56°C for 30 minutes. Remove from water bath and keep at room temperature. Appearance of red colour shows lactose is present. On standing, the colour increases.

Note

- The only sugars, which give a red colour, are disaccharides (lactose, maltose).
- Glucose, fructose, galactose, xylose and sucrose in large quantities give yellow colour, which is easily distinguished.

1.9. Osazone Test

Principle

When a solution of reducing sugars is heated with phenyl hydrazine, characteristic yellow crystalline compounds called osazones are formed. Because of similarities in the molecular structure, glucose, fructose and mannose produce same osazones.

Glassware and Equipments

Test tubes, test tube stand, pipettes, beakers, funnel, dropper, microscope

Reagents

1. Glacial acetic acid
2. Phenyl hydrazine hydrochloride
3. Sodium acetate

Procedure

Take 5 ml of sugar solution. Add 10 drops of glacial acetic acid. Add a pinch of phenyl hydrazine hydrochloride and double the amount of sodium acetate crystals. Mix and warm a little to see solids are dissolved. Filter the solution in another test tube and keep filtrate in boiling water bath for 20 min. Allow the tubes to cool slowly and examine the crystals under microscope.

Table 3. Examination of crystals

Crystal	Time for crystallization (min)	Shape of crystal
Glucosazone	5	Broomstick like
Fructosazone	2	Broomstick like
Galactosazone	7	Rhombic like
Lactosazone	10–12	Powder puff like
Maltosazone	10–15	Sunflower like

Note

- Sucrose does not produce osazones but on heating in boiling water bath for 30 min, get hydrolyzed and form osazones.

1.10. Tes–tape Test for Glucose

Principle

A commercially–available test paper is used for this test. This paper is impregnated with the enzymes; glucose oxidase and peroxidase, with an oxidizable substrate; orthotolidine. In the presence of glucose and moisture, the glucose oxidase will react with glucose to form H_2O_2 and gluconic acid. H_2O_2 formed is then acted upon by peroxidase to liberate O_2, which in turn acts upon the orthotolidine to give colour ranging from light green to bluish black, depending upon glucose concentration. This test is very sensitive and specific for glucose.

Glassware and Equipments

Test tubes, test tube stand

Reagents

1. Tes–tape papers
2. Sugar solution

Procedure

Moisten a small piece of Tes–tape with the solution to be tested. After about 1 minute, compare the darkest colour, which develops with the standard colours.

1.11. Identification Scheme for carbohydrates

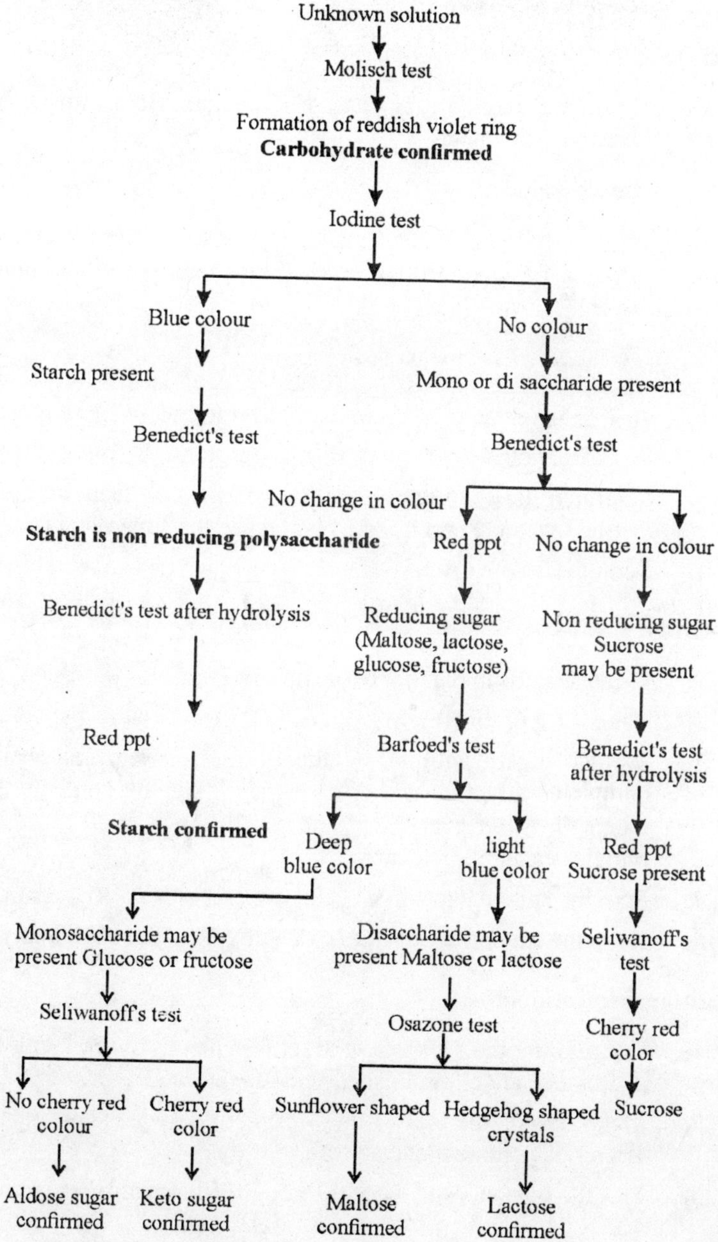

Quantitative Analysis

1.12. Extraction of Sugars

Principle

The free sugars can be extracted in 80% ethanol. Remaining residue is starch and other components.

Glassware and Equipments

Balance, test tubes, test tube stand, pipettes, measuring cylinder, beaker, volumetric flask, centrifuge tubes, centrifuge, pestle and mortar, flash evaporator

Reagents

1. 70% Ethanol: Add 30 ml of distilled water to 70 ml of ethanol.
2. 80% Ethanol: Add 20 ml of distilled water to 80 ml of ethanol.
3. Saturated lead acetate: Dissolve excess of lead acetate in distilled water. Keep it for 24 h and use the supernatant.
4. Sodium oxalate crystals.

Procedure

1. Grind 1 g of sample in a pestle and mortar.
2. Take 0.1 g of this finely ground sample in a centrifuge tube.
3. Extract sugars with 80% ethanol (hot) twice, followed by complete extraction with 70% ethanol, stir the contents using magnetic stirrer every time and centrifuge at 3000 rpm for 15 minutes each.
4. Pool the supernatants in a beaker.
5. Keep the sugar free residue in deep freezer for starch analysis.

Protein precipitation

1. Concentrate the alcoholic extract to aqueous syrup using flash evaporator under diminished pressure at 40°C.
2. Make its volume to 100 ml with distilled water and then add 1.5 ml of saturated lead acetate solution.
3. Mix the contents and filter it through Whatman filter paper no. 40.

4. Add a few crystals of sodium oxalate to the filtrate so as to precipitate excess lead ions.

Note

- After filtration, do not add more water to make up the volume.
- Keep the filtrate in refrigerator for further analysis.

1.13. Determination of Total Sugars by Phenol Method

Principle

Concentrated H_2SO_4 dehydrates sugars and forms furfurals (in case of pentoses) and hydroxyl methyl furfurals (in case of hexoses), which then reacts with phenol to yield blue green coloured condensed product. The absorbance of the coloured product is read at 490 nm.

Glassware and Equipments

Balance, test tubes, test tube stand, pipettes, measuring cylinder, beakers, volumetric flasks, spectrophotometer

Reagents

1. 5% distilled Phenol: Dissolve 5 g distilled phenol in distilled water to make its volume to 100 ml.
2. Concentrated H_2SO_4
3. Standard glucose solution: Dissolve 100 mg glucose in distilled water and make its volume to 100 ml. This will serve as stock standard solution. From the stock, take 10 ml and make its volume to 100 ml with distilled water to get working solution of 100 μg/ml.

Procedure

1. Take 1 ml of sugar solution and add 1 ml of 5% distilled phenol and then 5 ml of concentrated H_2SO_4.
2. Mix the contents thoroughly.
3. Cool the contents in test tubes under tap water.
4. Run a blank using 1 ml of distilled water instead of sample.
5. The optical density of colour developed is taken at 490 nm.

6. Prepare a standard curve in the range of 20–100 μg using glucose as standard.

Note

- Distillate the phenol before use.
- Phenol reagent should be clear.
- This method could be used for determining sugars eluted from chromatography, as the solvents do not interfere in this test.

1.14. Determination of Total Sugars by Anthrone Method

Principle

Concentrated H_2SO_4 dehydrates the sugars to furfurals and hydroxyl methyl furfurals, which combines with anthrone (10-keto-9,· 10-dihydro anthracene) to form a bluish green complex, a condensation product, which can be read at 625nm.

Glassware and Equipments

Balance, test tubes, test tube stand, pipettes, measuring cylinder, beakers, volumetric flasks, water bath, spectrophotometer

Reagents

1. 0.2% Anthrone reagent - Dissolve 200 mg anthrone in distilled water and make the volume to 100 ml.
2. Concentrated H_2SO_4
3. Standard glucose solution - Dissolve 100 mg glucose in distilled water and make the volume to 100 ml. This will serve as stock standard solution. From the stock take 10 ml and make the volume to 100 ml with distilled water to get working solution of 100 μg/ml.

Procedure

1. Take 1 ml of dilute protein–free sugar solution.
2. Add 4 ml of anthrone reagent.
3. Keep the tubes in boiling water bath for 10 minutes.
4. Remove the tubes and cool immediately under running tap water.

5. Similarly, prepare a blank using 1 ml of distilled water instead of sample.
6. Read the absorbance at 625 nm.
7. Prepare a standard curve in the range of 20–100 μg using glucose as standard.

Note

- This test is highly sensitive and the results depend on the time precision for boiling.
- This method does not suit for sugar samples extracted from paper chromatography, as even traces of organic solvents interfere in this method.
- As furfural forms less stable complex than hydroxymethyl furfural with anthrone reagent, this does not suit samples containing large amount of pentoses and is more sensitive for hexoses.
- Blank should be prepared carefully in duplicate.

1.15. Determination of Glucose content

Principle

In the presence of glucose and moisture, the enzyme glucose oxidase will react with glucose to form H_2O_2 and gluconic acid. H_2O_2 formed is then acted upon by peroxidase to liberate O_2, which in turn acts upon with o-dianisidine to give colour ranging from light green to bluish black, depending upon glucose concentration. This test is very sensitive and specific for glucose.

Glassware and Equipments

Balance, test tubes, test tube stand, pipettes, measuring cylinder, beakers, volumetric flasks, water bath, spectrophotometer

Reagents

1. Reagent A - Dissolve 30 mg of glucose oxidase [4 units/mg], 1.5 mg peroxidase [100 units/mg] in 5 ml of 0.1 M sodium phosphate buffer [pH = 7.0]

2. Reagent B - Dissolve 30 mg of o-dianisidine in 5 ml of ethanol. Filter the solution and store the filtrate in a refrigerator in an amber–coloured bottle.
3. Reagent C - Mix 5 ml of reagent A and 2.5 ml of reagent B. To it add 42.5 ml of 45% glycerol and store in a dark brown bottle in a deep freezer.
4. Standard glucose solution - Dissolve 1 g (10 mg/ml) glucose in distilled water and make the volume to 100 ml. This will serve as stock standard solution. From the stock take 10 ml and make the volume to 100 ml with distilled water to get working solution of 1 mg/ml.

Procedure

1. Take 1 ml of the sugar extract, and add 1 ml of reagent C in it.
2. Incubate the tubes at 30°C.
3. After incubation for 45–90 minutes, add 2 ml of 2 N HCl.
4. Read the colour absorbance at 540 nm.
5. Prepare a standard curve in the range of 0.1–1.0 mg using glucose as standard.

1.16. Determination of Total Reducing Sugars

Principle

Sugars containing free or potentially free aldehyde or ketone groups will reduce metal ions in an alkaline medium. Under careful conditions the amount of reduction is proportional to amount of reducing sugar. The sugar is heated with an alkaline solution of cupric tartarate. Cupric ions are reduced to cuprous oxide and which in turn reduces the arsenomolybdic acid to give molybdenum blue. The blue colour developed is determined spectrophotometrically at 620 nm.

Glassware and Equipments

Balance, test tubes, test tube stand, pipettes, measuring cylinder, beakers, volumetric flasks, water condensers, water bath, cyclomixer, spectrophotometer

Reagents

1. Reagent A - Dissolve 25 g anhydrous sodium carbonate, 25 g sodium potassium tartarate, 20 g sodium bicarbonate and 200 g

of anhydrous sodium sulphate in distilled water making final volume to 1 litre.

2. Reagent B - Dissolve 15 g copper sulphate in distilled water and make volume to 100 ml. Add 2–3 drops of concentrated H_2SO_4.
3. Reagent C - Mix afresh reagent A and B in 25:1 ratio.
4. Reagent D (Nelson reagent) - Dissolve 25 g ammonium molybedate in 450 ml water; add 21 ml concentrated H_2SO_4 slowly after stirring. Separately, dissolve 3 g sodium arsenate in 25 ml distilled water and add this solution drop wise to ammonium molybdate solution. Incubate the solution, so prepared for 2 h, at 37°C and store in an amber coloured bottle.
5. Standard glucose solution - Dissolve 100 mg glucose in distilled water and make the volume to 100 ml. This will serve as stock standard solution. From the stock take 10 ml and make the volume to 100 ml with distilled water to get working solution of 100 μg/ml.

Procedure

1. To 1 ml of appropriately diluted sugar solution, add 1 ml of freshly prepared reagent C.
2. Cover the tubes with water condensers and then keep in boiling water bath for 20 minutes.
3. After cooling the tubes to room temperature, add 1 ml of Nelson reagent (reagent D) to each tube and then add 7 ml of distilled water.
4. Mix the contents of the tubes thoroughly on a cyclomixer.
5. Read the absorbance at 510 nm against the reagent blank.
6. Prepare a standard curve in the range of 50–100 μg using glucose as standard.

1.17. Determination of Non-reducing Sugars

The difference in the concentration of total soluble and reducing sugars are taken as the concentration of non–reducing sugars.

Total Soluble sugars = Reducing Sugars + Non–reducing sugars

Non Reducing sugars [like sucrose] = Total Soluble sugars – Reducing Sugars

1.18. Determination of Fructose content

Principle

Ketohexoses react with acids much more rapidly than aldohexoses. Hydroxymethyl furfural reacts further with resorcinol to form brownish red–coloured complex, which can be read at 540 nm.

Glassware and Equipments

Balance, test tubes, test tube stand, pipettes, measuring cylinder, beakers, volumetric flasks, water condensers, water bath, spectrophotometer

Reagents

1. 1% Resorcinol - Dissolve 1 g resorcinol in 95% ethanol to make the volume to 100 ml.
2. 30% HCl - 100 ml of concentrated HCl diluted to 120 ml with distilled water.
3. Standard fructose solution - Dissolve 100 mg fructose in distilled water and make the volume to 50 ml. This will serve as stock standard solution. From the stock take 10 ml and make the volume to 100 ml with distilled water to get working solution of 200 μg/ml.

Procedure

1. Take 1 ml of sample sugar solution and add 1 ml of 1% resorcinol reagent.
2. To the contents add 3 ml of 30% HCl.
3. Heat the test tubes fitted with water condensers in a water bath at 80°C for 20 minutes.
4. Cool the tubes and read the absorbance at 490 nm.
5. Prepare a standard curve of fructose in the range of 10–200 μg.

1.19. Determination of Sucrose Content

Principle

Sucrose is a non–reducing disaccharide. It is first hydrolyzed to glucose and fructose (both are reducing). The hydrolysis can be carried out by heating the sample with HCl, either released fructose or glucose can be estimated. Present method estimates fructose.

Glassware and Equipments

Balance, test tubes, test tube stand, pipettes, measuring cylinder, beakers, volumetric flasks, water bath, spectrophotometer

Reagents

1. 1% Resorcinol solution - Dissolve 1 g of resorcinol in glacial acetic acid to make the volume to 100 ml.
2. 6% KOH solution - Dissolve 6 g KOH solution in distilled water and make the volume to 100 ml.
3. 30% HCl - 100 ml of concentrated HCl is diluted to 120 ml with distilled water.
4. Standard sucrose solution - Dissolve 100 mg sucrose in distilled water and make the volume to 100 ml. This will serve as stock standard solution.. From the stock take 10 ml and make the volume to 100 ml with distilled water to get working solution of 100 μg/ml.

Procedure

1. To 0.5 ml of properly diluted test sample, add 0.5 ml of 6% KOH solution.
2. Heat the tubes in a boiling water bath for 20 minutes to destroy free fructose.
3. Cool the tubes to room temperature, add 1% resorcinol solution and 3 ml of 30% HCl.
4. Incubate the tubes at 80°C for 10 minutes
5. Read the intensity of the resulting pink colour developed at 490 nm.
6. Prepare a standard curve in the range of 50–100 μg using sucrose as standard.

Note

- The determined values may include minor content of fructose bound in other forms, e.g., glucofructosans, stachyose, raffinose etc.

1.20. Determination of Starch Content

Principle

The starch gets hydrolyzed to simple sugars in the presence of perchloric acid. After hydrolysis free sugars can be estimated by phenol or anthrone method. The concentration of sugars obtained is multiplied with a factor of 0.9 to calculate the concentration of starch.

Glass ware and Equipments

Test tubes, test tube stand, pipettes, measuring cylinder, beakers, volumetric flasks, funnel, magnetic stirrer and beads, centrifuge, centrifuge tubes, spectrophotometer

Reagents

1. 52% Perchloric acid - Dissolve 52 g of perchloric acid to make the volume to 100 ml.
2. 5% distilled Phenol - Dissolve 5 g of distilled phenol to make the volume to 100 ml.
3. Concentrated H_2SO_4
4. Standard glucose solution (100 μg/ml) - Dissolve 100 mg glucose in distilled water and make the volume to 100 ml. This will serve as stock standard solution. From the stock take 10 ml and make the volume to 100 ml with distilled water to get working solution of 100 μg/ml.

Procedure

1. To the sugar free residue (obtained as described in 1.12), add 5 ml of water followed by 6.5 ml of perchloric acid.
2. Stir on magnetic stirrer for 5 minutes and then intermittently for another 15 minutes.
3. Add 20 ml of distilled water and centrifuge the contents at 3000 rpm for 10 minutes.
4. Decant the supernatant in a 100 ml volumetric flask.
5. Repeat the above process with the remaining residue.
6. Pool the supernatants and make the volume to 100 ml with distilled water.
7. Filter through Whatman filter paper no.1 and discard the first few ml.

8. Take 1 ml of sample, add 1 ml of 5% phenol and then 5 ml of concentrated H_2SO_4 and mix the contents thoroughly.
9. Cool the contents and read the optical density at 490 nm.
10. A standard curve of glucose solution is prepared in the concentration range from 20–100 μg.
11. Calculate the glucose concentration. Starch concentration is calculated from glucose concentration by multiplying with a factor of 0.9.

2 AMINO ACIDS AND PROTEINS

PROTEINS

Proteins are macromolecules made up of different amino acids. Amino acids are linked together in a polypeptide chain through peptide or amide bonds. So, proteins are unbranched polymers of L–α–amino acids. The general formula of L–α–amino acid is given as:

$$\begin{array}{c} R \\ | \\ HOOC - C - NH_2 \\ | \\ H \end{array}$$

R - represents side chain which is variable.

There are 20 amino acids: Glycine, alanine, valine, leucine, isoleucine, proline, asparagine, glutamine are neutral amino acids. Serine and threonine are hydroxyl group–containing amino acids. Phenylalanine, tryptophan and tyrosine are aromatic amino acids. Cysteine, methionine are sulphur–containing amino acids. Aspartate and glutamate are acidic in nature and histidine, arginine, lysine are basic in nature.

Proteins can be classified in three ways.

1. Based on their molecular structure, they are of two types

a. Globular proteins: These proteins have their peptide chains in compactly folded form e.g., albumins and globulins.

b. Fibrous proteins: These proteins are fibre like in structure e.g., keratins and collagen.

2. On the basis of their functions, proteins may be classified as:

Table 4. Classification of proteins on the basis of their function

Sr. No.	Protein	Example
1.	Carrier proteins	Transferrin

2.	Contractile proteins	Actin myosin
3.	Genetic proteins	nucleo proteins
4.	Defense proteins	Immunoglobulins (Ig)
5.	Enzyme proteins	Pepsin, trypsin
6.	Respiratory proteins	Cytochromes, haemoglobin
7.	Storage proteins	Albumins, globulins
8.	Structural proteins	Keratin, collagen
9.	Transport proteins	hormones

3. Proteins may be classified in another way as

a. Simple proteins

These include albumins, globulins, glutilins, prolamines, histones, globins, protamines and albuminoids.

b. Compound or conjugated proteins

These include nucleoproteins, phosphoproteins, glycoproteins, chromoproteins, lipoproteins and metaloproteins.

c. Derived proteins

i. Primary derived proteins

- Proteans are insoluble products formed from protein by the action of water, dilute acids and enzymes e.g., fibrin.
- Metproteins are formed by the action of concentrated acids and alkalies on proteins.

ii. Secondary derived proteins

These are obtained by progressive hydrolysis of proteins and include proteoses, peptones and peptides.

Qualitative Analysis

2.1. Biuret Test

Principle

Proteins form a violet or purple coloured complex with cupric ions in alkaline medium. In this case proteins react in the same way as the biuret compound reacts with cupric ions. This test is positive for all compounds containing more than one peptide linkage, e.g., proteins and their hydrolytic products (metaproteins, proteoses, peptones, polypeptides except dipeptides and amino acids). The substances, which contain two carbamyl (-$CONH_2$) groups joined either directly or through a single atom of carbon or nitrogen, and similar substances, which contain –$CSNH_2$,- C(NH) NH_2 or CH_2NH_2 group also respond to this biuret test. Hence, oxamide ($CONH_2$-$CONH_2$) and biuret ($CONH_2$-NH-$CONH_2$) give test positive. Histidine amino acid also gives this test positive.

Glassware and Equipments

Test tubes, test tube stand, pipettes, beakers, dropper, electric balance, reagent bottles, volumetric flasks

Reagents

1. 1% $CuSO_4$: Dissolve 1 g $CuSO_4$ in distilled water and make its volume to 100 ml.
2. 10 N NaOH: Dissolve 40 g NaOH in distilled water and make its volume to 100 ml.

Procedure

Take 1 ml protein solution and add 1 ml of 10 N NaOH solution to it. Mix the contents. Then slowly add 1 or 2 drops of 1% $CuSO_4$ solution. Formation of violet colour shows presence of peptide linkage in the molecules.

Note

- If more drops of $CuSO_4$ are added, blue colour develops instead of violet.

2.2. Ninhydrin (Triketohydrindene Hydrate) Test

Principle

Ninhydrin (triketohydrindene hydrate),when heated with free amino acids, gets reduced and amino acid is oxidized to an aldehyde with one carbon less than original, and a molecule of CO_2 is released. In turn, the reduced ninhydrin is condensed with another molecule of ninhydrin in the presence of NH_3 resulting in the formation of a purple coloured complex. This test is positive for proteins, peptones, peptides and amino acids. It is also positive with other primary amines including NH_3.

Secondary amines (proline and hydroxyproline) also react with ninhydrin but, give yellow colour.

Glassware and Equipments

Test tubes, test tube stand, pipettes, beakers, dropper, volumetric flasks, electric balance, reagent bottles, Bunsen burner, tongs

Reagents

1. 0.2% ninhydrin solution: Dissolve 200 mg ninhydrin in distilled water to make its volume to 100 ml.

Procedure

Take 1 ml of protein solution and add two drops of freshly prepared 0.2% ninhydrin solution to it. Mix the contents and then boil it. Development of blue colour shows presence of proteins, peptides or amino acids.

2.3. Millon's Test for Hydroxybenzene Ring (Tyrosine)

Principle

The neutral protein solution reacts with acidified $HgCl_2$ to form yellow precipitate of Hg- protein complex. This complex forms a nitrite complex with sodium nitrite ($NaNO_2$) solution on heating and develops a red colour. The test is positive for phenols or phenol substances such as salicylic acid. Tyrosine is only amino acid, which shows this test positive.

Glassware and Equipments

Test tubes, test tube stand, pipettes, electric balance, reagent bottles, beakers, dropper, volumetric flasks, Bunsen burner, tongs

Reagents

1. Millon reagent: 10% $HgSO_4$ in 10% H_2SO_4
2. 1% sodium nitrite: Dissolve 1 g sodium nitrite ($NaNO_2$) in distilled water to make its volume to 100 ml.

Procedure

Take 1 ml of protein solution and add 1 ml of Millon's reagent to it. Mix the contents and boil it. Formation of yellow precipitate takes place.

Cool under the tap and add a drop of 1% sodium nitrite solution. On heating red colour develops, showing presence of hydroxybenzene group.

Note

- Excess of Cl^- interferes by forming unionized $HgCl_2$. So, add more $HgSO_4$.

2.4. Xanthoproteic Reaction or Yellow Protein Test (For tyrosine, phenylalanine and tryptophan)

Principle

Proteins in solution form white precipitate on addition of concentrated HNO_3, which turns yellow on heating and then becomes orange in alkaline medium. Insoluble proteins (say skin) turn yellow and orange on the surface. This reaction is given by aromatic amino acids such as tyrosine, phenylalanine and tryptophan.

Glassware and Equipments

Balance, test tubes, test tube stand, pipettes, electric balance, reagent bottles, beakers, volumetric flasks, Bunsen burner, tongs

Reagents

1. Concentrated HNO_3
2. 10 N NaOH: Dissolve 40 g NaOH in distilled water to make its volume to 100 ml.

Procedure

Take 1 ml of protein solution and add 1 ml of concentrated HNO_3 to it. White precipitates are formed. Boil it. On boiling, colour changes to yellow. Add 2 ml of 10 N NaOH to make it alkaline. Colour changes to orange, indicating presence of aromatic amino acids.

2.5. Hopkins–Cole–Adam–Kicwicz Reaction (For Tryptophan)

Principle

Tryptophan (amino acid with indole ring) in protein forms a condensation product with the formaldehyde or glyoxylic acid (present in glacial acetic acid, which is exposed to light) in the presence of concentrated H_2SO_4 and develops a purple colour. The colour is formed as a result of the reaction between the indole nuclei of amino acid and aldehyde.

Glassware and Equipments

Balance test tubes, test tube stand, pipettes, reagent bottles, beakers, volumetric flasks

Reagents

1. Glacial acetic acid (Expose to light before use)
2. H_2SO_4

Procedure

Take 1 ml of protein solution. Add 2 ml of glacial acetic acid and then add 1 ml concentrated H_2SO_4 by the side of tube. The appearance of purple ring at the junction of two liquids indicates presence of tryptophan.

2.6. Sakaguchi Reaction

Principle

Arginine (free or in bound form in proteins) in the presence of sodium hypochlorite forms a complex with alcoholic α–napthol and slowly a bright red colour is developed. Different substances containing guanido group gives this test positive. As arginine is the only amino acid with guanido group in proteins, so this test is specific for arginine.

Glassware and Equipments

Balance, test tubes, test tube stand, pipettes, electric balance, reagent bottles, beakers, volumetric flasks, dropper

Reagents

1. 5% α–napthol: Dissolve 5 g of α–napthol in 95% alcohol to make its volume to 100 ml.
2. 10 N NaOH: Dissolve 40 g NaOH in distilled water to make its volume to 100 ml.
3. 10% sodium hypochlorite (NaOCl): Dissolve 10 g NaOCl in distilled water to make its volume to 100 ml.

Procedure

Take 1 ml of protein solution, and add 1 ml of 10 N NaOH and 2 drops of 5% α–napthol to it. Mix the contents and then add 1 ml of 10% sodium hypochlorite solution. Shake and note the colour. Development of bright red colour shows presence of arginine.

Note

- This test is highly sensitive and may be used as general test for proteins because all known proteins contain sufficient arginine.

2.7. Sulphur Test

Principle

Sulphur of the sulphur–containing amino acids (cysteine, cystine and methionine) reacts with sodium hydroxide forming sodium sulphide (Na_2S).When lead acetate is added, a black or brown precipitate of lead sulphide (PbS) is formed as a result of the reaction between sodium sulphide and lead acetate. This PbS is insoluble in dilute HCl.

Glassware and Equipments

Balance, test tubes, test tube stand, pipettes, electric balance, reagent bottles, beakers, volumetric flasks, Bunsen burner, tongs

Reagents

1. 10 N NaOH: Dissolve 40 g NaOH in distilled water to make its volume to 100 ml.
2. 5% lead acetate: Dissolve 5 g lead acetate in distilled water to make its volume to 100 ml.

Procedure

Take 1 ml of protein solution and add 1 ml of 10 N NaOH. Mix the contents and boil for 1 minute. Then, add 3 ml of 5% lead acetate. A black or brown precipitate is formed showing that the test is positive for sulphur amino acid.

2.8. Coagulation Reaction

Principle

Coagulation of protein is caused by the denaturation of protein structure by heat or acids. Denaturation refers to breakdown of hydrogen bonds of secondary and tertiary protein structure while the primary structure remains unaffected.

Glassware and Equipments

Test tubes, test tube stand, pipettes, beakers, volumetric flasks, Bunsen burner, tongs

Reagents

1. 2% Acetic acid: Take 2 ml of acetic acid and make its volume to 100 ml with distilled water.
2. HNO_3

Procedure

1. Take 2/3 part protein solution in a test tube. Heat upper portion of the solution. An opalescent appears, which becomes deep on addition of a few drops of 2% acetic acid. This shows presence of albumin.
2. Coagulation with concentrated HNO_3: Take 2 ml of concentrated HNO_3 and add protein solution to it by the side of tube. A white ring appears at the junction of two liquids.

2.9. Precipitation Reaction

2.9.1. Precipitation by using Salts of Heavy Metals

Principle

Precipitation of protein is caused by using salts of heavy metals such as $HgCl_2$, $AgNO_3$, $CuSO_4$ and lead acetate. Heavy metals (e.g., Hg^{2+}, Pb^{2+}, Cu^{2+}) are high molecular weight cations. The positive charge of these cations counteracts the negative charge of the carboxylate group in proteins and hence precipitates the proteins.

Glassware and Equipments

Balance, test tubes, test tube stand, pipettes, beakers, volumetric flasks

Reagents

1. 1% $ZnSO_4$: Dissolve 1 g $ZnSO_4$ in distilled water to make its volume to 100 ml.
2. 0.5% $FeCl_3$: Dissolve 0.5 g $FeCl_3$ in distilled water to make its volume to 100 ml.

Procedure

1. Take 2 ml of dilute protein solution (egg white) and add a few drops of $ZnSO_4$ (1%) drop wise. White precipitates are formed.
2. Take 2 ml of dilute protein solution (egg white) and add a few drops of 0.5% $FeCl_3$ solution drop wise. An opalescent appears and disappears on addition of excess of $FeCl_3$ solution.

2.9.2. Precipitation by Alkaloid Reagent

Principle

Proteins are precipitated by certain acids, some of which are called alkaloid reagents (picric acid, phosphotungstic acid, tannic acid, meta-phosphoric acid). Alkaloidal reagents are high molecular weight anions. The negative charge of these anions counteracts the positive charge of the amino group in proteins giving a precipitate.

Glassware and Equipments

Test tubes, test tube stand, pipettes, beakers, volumetric flasks

Reagents

1. 2% Sulphosalicylic acid - 2% solution of salicylic acid in H_2SO_4.
2. Esbach's reagent - Solution of picric and citric acid.

Procedure

1. Take 1 ml of protein solution and add 1 ml of Esbach's reagent (solution of picric and citric acid) to it. Yellow precipitates are formed.
2. Take 1 ml of protein solution and add a few drops of 2% solution of sulphosalicylic acid to it. White precipitates are formed.

2.9.3. Precipitation by Neutral Salts

Principle

Proteins are precipitated by concentrated solutions of NaCl, $(NH_4)_2SO_4$ and Na_2SO_4. These salts precipitate proteins by salting out.

Glassware and Equipments

Test tubes, test tube stand, pipettes, beakers

Reagents

1. Saturated $(NH_4)_2SO_4$: Dissolve excess of ammonium sulphate in water and keep it overnight. Take the supernatant and discard the crystals settled at the bottom.

Procedure

1. Full saturation: Take 5 ml of protein solution. Add solid $(NH_4)_2SO_4$ till it becomes saturated. Formation of gelatinous precipitates indicates presence of albumin.
2. Half Saturation: Take 3 ml of protein solution. Add 3 ml of saturated $(NH_4)_2SO_4$ solution. Shake vigorously. A gelatinous precipitates shows presence of globulin.

2.9.4. Precipitation by Dehydrating Agents: Alcohol, Acetone etc.

Principle

These agents convert proteins into suspensoids, which flocculate upon the addition of a few drops of salt solution. Alcohol also causes denaturation of proteins by bringing the proteins into isoelectric point at which it is precipitated.

Glassware and Equipments

Test tubes, test tube stand, pipettes, beakers

Reagents

1. Alcohol

Procedure

1. Take 1 ml of protein solution. Add 2–3 ml of alcohol. A white opalescent formed indicates the presence of protein. Filter after 10 min. Try to dissolve precipitates in water, they will not dissolve.
2. Take 1 ml of protein solution. Take 2–3 ml of alcohol in another tube. Place both tubes in the freezer for 10 min. Then mix the contents of the two tubes. A white opalescent appears. Filter the contents. Dissolve the residue on filter paper and observe that the precipitate dissolves.

2.9.5. Precipitation at Isoelectric Point

Principle

Isoelectric point is the pH at which a protein has no net charge and gets precipitated.

Glassware and Equipments

Test tubes, test tube stand, pipettes, dropper, volumetric flasks, Bunsen burner, tongs

Reagents

1. 2% acetic acid: Dissolve 2 ml acetic acid in distilled water to make its volume to 100 ml.
2. Bromocresol green

Procedure

Take 3 ml of protein solution. Add a drop of bromocresol green. A blue colour develops. Then add 2% acetic acid drop wise till colour becomes green, at 4.7 pH. A flocculent precipitate of casein protein is formed. Boil it, cool it and add dilute NaOH. Precipitate dissolves.

Quantitative Analysis

2.10. Determination of Total Soluble Protein

Principle

This is widely used method of protein estimation. It is much more sensitive than Biuret method. In an alkaline solution, protein and copper forms a complex. This complex reduces the phosphomolybdate–phosphotungstate reagent (Folin Ciocalteau reagent) forming a blue coloured complex. The intensity of blue colour developed is read at 520 nm.

Glassware and Equipments

Balance, test tubes, test tube stand, pipettes, measuring cylinder, beakers, volumetric flasks, vortex mixer, spectrophotometer

Reagents

1. Solution A: 2% Sodium carbonate (Na_2CO_3) in 0.1N sodium hydroxide. Dissolve 4g NaOH in water and make its volume to 100 ml. It gives 0.1 N NaOH. To it, add 2 g sodium carbonate.

2. Solution B: Prepare a fresh solution by mixing equal volume of (a) and (b).
    a. 1% Copper sulphate ($CuSO_4.5H_2O$):1g of copper sulphate is dissolved in 100 ml distilled water.
    b. 2% Sodium potassium tartarate solution: 2 g Sodium potassium tartarate is dissolved in 100 ml distilled water.
3. Solution C (Alkaline copper solution): Mix 50:1 parts of reagent A and reagent B, and prepare freshly.
4. Solution D: Folin Ciocalteu Reagent (2N). Dilute with an equal volume of water (1:1 ratio) before use.
5. Bovine serum albumin (BSA): Dissolve 250 mg BSA in distilled water and make the volume to 100 ml. This will serve as stock standard solution. From the stock take 10 ml and make the volume to 100 ml with distilled water to get working solution of 250μg/ml.

Procedure

1. Take 1 ml of suitably diluted protein extract.
2. Add 5 ml of reagent C.
3. After 10 min, add 0.5 ml of Folin- phenol reagent D.
4. Mix the contents by vortex mixer.
5. Keep the tubes for 30 min.
6. Read the absorbance of developed colour at 520 nm using spectrophotometer.
7. Prepare a standard curve by taking Bovin serum albumin as a standard protein in the range of 50–250 μg.
8. Calculate the amount of protein in mg/g or kg 100 g sample from the standard curve.

2.11. Determination of Total Protein by Biuret Method

Principle

Proteins form a violet/purple–coloured complex with cupric ions in alkaline medium. In this case, proteins react in the same way as the Biuret compounds react with cupric ions. The intensity of purple colour is read at 540 nm photometrically.

Glassware and Equipments

Balance, test tubes, test tube stand, pipettes, measuring cylinder, beakers, volumetric flasks, vortex mixer, spectrophotometer

Reagents

1. Biuret reagent - Dissolve 4 g NaOH in distilled water. To it add 4.5 g of sodium potassium tartarate and mix the contents. Add 1.5 g of $CuSO_4$ and 4.5 g KI. Transfer the contents in a volumetric flask of 500 ml and make up the volume with distilled water.
2. Bovine serum albumin (BSA) - Dissolve 250 mg BSA in distilled water and make the volume to 100 ml. This will serve as stock standard solution. From the stock take 10 ml and make the volume to 100 ml with distilled water to get working solution of 250 μg/ml.

Procedure

1. Take 1 ml of suitably diluted protein extract.
2. To it add 3 ml of Biuret reagent.
3. Mix the contents by vortex mixer.
4. Read the absorbance of developed colour at 540 nm using spectrophotometer.
5. Prepare a standard curve by taking Bovine serum albumin (BSA) as a standard protein in the range of 50–250 μg.
6. Calculate the amount of protein in sample from the standard curve and express the results as mg/g or mg/100 g.

2.12. Determination of Total Protein by Bradford's method

Principle

Dye coomassie brilliant blue G-25 binds to the free amino group of proteins, as a result absorption maxima of dye changes from 465 to 595 nm. This method is very sensitive. It is a relative method as the amount of dye binding appears to vary with the content of basic amino acid residues in the protein.

Glassware and Equipments

Balance, test tubes, test tube stand, pipettes, measuring cylinder, beakers, volumetric flasks, spectrophotometer

Reagents

1. Dye coomassie brilliant blue G-25 - Dissolve 10 mg of dye in 5 ml of absolute ethanol, add 10 ml of 85 % H_3PO_4, to this add 185 ml of distilled water and filter it.
2. Bovine serum albumin (BSA) - Dissolve 250 mg BSA in distilled water and make the volume to 100 ml. This will serve as stock standard solution. From the stock take 10 ml and make the volume to 100 ml with distilled water to get working solution of 250 μg/ml.

Procedure

1. Take 50 μl of protein sample.
2. Add 5 ml of coomassie brilliant blue dye solution.
3. Incubate for 1 h at room temperature.
4. Read the absorbance at 595 nm.
5. Prepare a standard curve by taking Bovine serum albumin (BSA) as a standard protein in the range of 50–250 μg.
6. Calculate the amount of protein in sample from the standard curve and express the results as mg/g or mg/100 g.

Note

- Dye should be pure. If it is not pure, purify it by ammonium sulphate or ammonium acetate precipitation. Dye rarely dissolves completely even after purification. Filter the dye solution with Whatman filter paper no. 1.

2.13. Determination of Total Protein by Amido Black Filter Assay

Principle

Protein to be assayed is precipitated on a filter paper using trichloroacetic acid. The filter is then stained and destained with a protein stain. The stain is then eluted from the filter and binding is quantified spectrophotometrically.

Glassware and Equipments

Balance, test tubes, test tube stand, pipettes, measuring cylinder, beakers, volumetric flasks, funnel, vortex mixer, spectrophotometer

Reagents

1. Tris–SDS solution (1% SDS in 1M Tris HCl, pH 7.5) - Dissolve 1 g SDS in 1M Tris HCl of pH 7.5 and make its volume to 100 ml.
2. 60% Trichloroacetic acid - Dissolve 60 g TCA in distilled water and make the volume to 100 ml.
3. Staining solution [1% Amido black 10 B in methanol: acetic acid: water (45:10:45)] - Dissolve 1 g amido black in methanol: acetic acid: water (45:10:45) to make its volume to 100 ml.
4. Destaining solution: Methanol: acetic acid: water (90:2:18)
5. Eluant: 0.05 M EDTA, 25 mM NaOH in 50% ethanol
6. Nitrocellulose membrane filter
7. Bovine serum albumin (BSA) - Dissolve 250 mg BSA in distilled water and make the volume to 100 ml. This will serve as stock standard solution. From the stock take 10 ml and make the volume to 100 ml with distilled water to get working solution of 250 μg/ml.

Procedure

1. Take 50 μl of sample and dilute it to 270 μl.
2. Add 30 μl of Tris SDS solution, followed by 600 μl of 60% TCA, allow it to stand for 15 minutes.
3. Take pre–wet nitrocellulose membrane filter disc (25 mm in size with 0.45 μm pore size).
4. Filter the above solution through it, rinse the tubes twice with 2 ml of 60% TCA and use the rinse to wash the filter.
5. Remove the filter from the filtration apparatus and incubate the filter for 2–3 minutes in staining solution.
6. Rinse the filter with distilled water twice or thrice, followed by destaining solution and again with distilled water.
7. Blot dry excess water and incubate with eluant for 10 minutes and vortex it.
8. Add 0.9 ml of eluant, decant and read the absorbance at 630 nm.
9. Prepare a standard curve by taking BSA as a standard protein in the range of 10–250 μg.

10. Once stain has been eluted, it can be kept for long time before taking reading.

2.14. Determination of Crude Protein by Kjeldahl Method

Principle

During digestion with sulphuric acid (H_2SO_4) nitrogen of the protein of organic molecules is converted into ammonium sulphate. Ammonia is liberated from the ammonium sulphate during distillation. The liberated ammonia is collected in boric acid solution and titrated against standard hydrochloric acid/sulphuric acid.

Glassware and Equipments

Balance, Kjeldahl digestion and distillation apparatus, Kjeldahl tubes, burette, pipette, volumetric flasks, conical flasks

Reagents

1. Digestion mixture (K_2SO_4:$CuSO_4$) in ratio of 10:1
2. Concentrated H_2SO_4
3. N/10 HCl: Measure 8.47 ml of concentrated HCl (11.8 N) and dilute to 1000 ml with distilled water.
4. 2% Boric acid - Dissolve 2 g boric acid in distilled water and make the volume to 100 ml.
5. 40% NaOH - Dissolve 40 g NaOH in distilled water and make the volume to 100 ml.
6. Mixed indicator - Bromocresol green (0.5 g) and methyl red (0.1 g) dissolved in 100 ml of 95% alcohol and pH adjusted to 4.5 with HCl or NaOH.

Procedure

1. Take 100 mg of dry sample in to the digestion flask.
2. To it add 2 g of digestion mixture and 5 ml of concentrated H_2SO_4.
3. Heat the flask till the solution becomes clear.
4. Cool the digested contents and make volume to 25 ml in volumetric flasks with distilled water.
5. Distill an aliquot (5 ml) of this solution in microkjeldahl distillation apparatus by adding 4–10 ml of 40% NaOH.

6. The liberated NH_3 is trapped in 5 ml of boric acid to which 2–3 drops of mixed indicator are added.
7. Collect 30 ml of distillate in 100 ml of conical flask.
8. Titrate it with N/10 HCl till its colour changes from blue to faint pink.
9. Simultaneously run a blank without sample.

N content is calculated as:

$$\% \text{ N in sample} = [(T\text{-}B) \times N \times 1.4 \times V_1] / [V_2 \times S]$$

T - Sample titration

B - Blank titration

N - Normality of standard acid

S - Sample weight

V_1 -Total volume made

V_2 - volume used for distillation

Crude protein content (%) = N x 6.25

2.15. Estimation of Non–protein Nitrogen

Sample is extracted with cold 10% trichloroacetic acid. This leads to the precipitation of protein and extraction of non-protein nitrogen. Supernatant collected after centrifugation is diluted to a fixed volume (25 ml) and distilled as described in section 2.14 point 5 onwards.

Crude protein content (%) = N x 6.25

True protein (%) = (total nitrogen – non–protein nitrogen) x 6.25

2.16. NIRS (Near Infrared Spectroscopy) Method of protein estimation

This is a non–destructive method for estimation of protein content from the intact sample.

Principle

The Infrared radiation is heat radiation. The energy of this radiation corresponds to energy differences between different vibrational modes in molecules. The near infrared spectroscopy is based upon the interaction of near infrared radiation (part of the electromagnetic spectrum that is between the visible and mid-infrared regions; 700–2500 nm) with matter. The reflected radiation from the sample surface at a given wavelength is calculated by measurement of reflectance of a standard reference at same wavelength. With the help of computer software a calibration equation is developed, which is used to determine the concentration of protein in the given sample (Fig. 7).

Glassware and Equipments

Sample cup or vial, NIR (near infrared) spectrophotometer

Reagents

1. Samples

Procedure

1. A very small amount of sample is placed in sample cup.
2. It is scanned using NIR spectrophotometer.
3. Using already prepared calibration model, protein content is quantified by the standard software.

Fig. 7. Fourier Transformer Near Infrared Analyser

2.17. Determination of Free Amino Acid Content by Ninhydrin Method

Principle

Ninhydrin (triketohydrindene hydrate) reacts with α–amino acids to give a purple–coloured compound. The developed colour is read at 570 nm. All amino acids do not give exactly the same intensity of colour. The amino acids proline and hydroxyl proline give a yellow colour, which is read at 440 nm.

Glassware and Equipments

Balance, test tubes, test tube stand, pipettes, measuring cylinder, beakers, volumetric flasks, pestle and mortar, centrifuge tubes, centrifuge, water bath, spectrophotometer

Reagents

1. 0.1% ninhydrin in 0.5 M citrate buffer (pH 5.5) - Dissolve 100 mg ninhydrin in 0.5 M citrate buffer to make its volume to 100 ml.
2. Pure glycerol
3. 0.5 M citrate buffer: See annexure Table 1.5.
4. Ninhydrin reagent: Mix reagents 1, 2 and 3 in the ratio of 5:12:2.
5. L–glycine standard - Dissolve 100 mg glycine in distilled water and make the volume to 100 ml. This will serve as stock standard solution. From the stock take 10 ml and make the volume to 100 ml with distilled water to get working solution of 100 μg/ml.

Procedure

1. Take 100 mg of dried material.
2. Grind the sample with pestle and mortar and extract free amino acids with 3 ml of 80% ethanol, centrifuge the contents at 3000 x g.
3. Repeat the extraction twice with 80% ethanol, pool the supernatants and make the volume to 10 ml.
4. Take a known aliquot (0.5 ml) of extract in a test tube, make its volume to 1 ml with 80% ethanol.
5. To it add 5 ml of ninhydrin reagent.
6. Mix the contents gently and keep for 10–15 minutes in a boiling water bath.
7. Cool the tubes and take absorbance at 570 nm.
8. Similarly run a blank using 1 ml of 80% ethanol.

9. Prepare a standard curve by taking L–glycine as a standard amino acid in the range of 10–100 μg.

2.18. Determination of Free Proline Content

Principle

Proline is extracted in aqueous sulphosalicylic acid. The extracted proline reacts with ninhydrin under strong acidic conditions (pH 1.0) and intensity of red colour developed is read at 520 nm.

Glassware and Equipments

Balance, test tubes, test tube stand, pipettes, measuring cylinder, beakers, volumetric flasks, pestle and mortar, centrifuge tubes, centrifuge, water bath, ice bath, spectrophotometer

Reagents

1. Acidic ninhydrin - Dissolve 1.25 g of ninhydrin in 30 ml of glacial acetic acid and 20 ml of 6 M phosphoric acid.
2. 3% sulphosalicylic acid - Take 3 ml of sulphosalicylic acid and make final volume to 100 ml.
3. Glacial acetic acid
4. Toluene
5. L–proline standard - Dissolve 100 mg proline in distilled water and make the volume to 100 ml. This will serve as stock standard solution. From the stock take 10 ml and make the volume to 100 ml with distilled water to get working solution of 100 μg/ml.

Procedure

1. Take 100 mg of dried plant material.
2. Grind the sample with pestle and mortar and extract free amino acids with 3 ml of 3% sulphosalicylic acid, centrifuge the contents at 3000 x g for 15 minutes.
3. Repeat the extraction twice with 3% sulphosalicylic acid, pool the supernatants and make the volume to 10 ml.
4. Take a known aliquot (1.0 ml) of extract in a test tube.
5. To it add 2 ml of ninhydrin, and 2 ml of glacial acetic acid.

6. Mix the contents gently and keep for 1 hour in a boiling water bath.
7. Terminate the reaction by placing the tubes in ice bath.
8. Add 4 ml of toluene to the reaction mixture and shake the contents for a few seconds.
9. Separate the toluene layer and read the intensity of red colour developed at 520 nm.
10. Similarly run a blank using 1.0 ml of 3% sulphosalicylic acid.
11. Prepare a standard curve by taking L-proline as a standard amino acid in the range of 10–100 μg.
12. Calculate free proline content from given sample using standard curve.

3 HYDROGEN ION CONCENTRATION (pH) AND BUFFERS

The term pH was introduced by Sorenson in 1909. pH is defined as the negative log of H^+ ions. It is given as, pH= –log H^+.

The resistance of a solution to the changes in hydrogen ion concentration upon the addition of small amounts of acid or alkali is termed buffer action. The solution which possesses such properties is called a buffer solution.

A buffer is said to possess reserve acidity and reserve basicity. A buffer solution usually consists of solutions containing a weak acid [HA] and its sodium or potassium salt [A^-] or a weak base B and its salt [BH^+].

3.1. Determination of pH of a Buffer

Principle

pH can be measured by the use of pH paper, indicator dyes, with use of pH meter. Use of pH meter is the most convenient and reliable method. pH meter measures the electromotive force of a reference electrode, the test solution and a glass electrode sensitive to hydrogen ions or it measures the electro motive force (emf) arising due to the potential difference between hydrogen reference electrode and electrode containing test samples.

Glassware and Equipments

pH meter, pipettes, beakers, reagent bottles

Reagents

1. Standard buffers of pH 4.0, 7.0 and 9.0

Procedure

1. Dip the electrode in standard solution.

2. Calibrate the pH meter with standard buffer of pH 7.0 and 4.0 or 9.0 or both.
3. Once calibration is over, wash the electrode and dip it in the test solution, and note the reading.

Note

- Wash the electrode before and after every use.
- If electrode is new, soak it in 0.1M HCl or distilled water for several hours before use.
- Ensure uniform mixing of solution before measuring pH.
- All the buffers to be used should be taken out of refrigerator and brought to room temperature since k_w, pk_a and pH of buffer vary with temperature.
- Glass electrode remains active down to quite low pH values, but in strong alkali recorded pH is often too low because of interference by other ions such as Na^+. So, ordinary glass electrode should not be used for pH above 10.
- Always dip the electrode in distilled water and never let it dry.

3.2. Preparation of a Buffer (100 ml) of pH 4.0, given is pk_a of 4.73

Principle

Buffer is a solution that resists change in pH on addition of acid or alkali. Henderson Hasselbalch equation can be used to find relative amount of two components required as:

$$pH = pk_a + \log [A-]/[HA]$$

Maximum buffering capacity is shown when pH = pk_a. If a buffer shows minimum change, it means that it has maximum buffer value.

Glassware and Equipments

pH meter, pipettes, beakers, measuring cylinders

Reagents

1. 0.2 M sodium acetate : Dissolve 16.4 g of $C_2H_3O_2$ Na or 27.2 g $C_2H_3O_2$ Na $3H_2O$ in distilled water and make volume to 1 litre.
2. 0.2 M acetic acid: Take 11.15 ml of acetic acid and make volume to 1 litre.

Procedure

1. Using Henderson Hasselbalch equation, find the amount of components.
2. Take the acid component in a beaker.
3. Add salt component slowly with constant stirring.
4. Set pH to 4.0.

Calculations

Henderson Hasselbalch equation

$$pH = pk_a + \log [A^-] / [HA]$$

Given pH = 4.0 and $pk_a = 4.73$

Putting values, we get,

$$4.0 = 4.73 + \log [A-] / [HA]$$

$$\text{Log } [A^-] / [HA] = 4.0 - 4.73$$

$$\text{Log } [A^-] / [HA] = -0.73$$

$$\text{Log } [HA] / [A^-] = 0.73$$

Taking antilog 0.73 = 5.4, we get,

$$[HA] / [A^-] = 5.4/1.0$$

$$\text{Acid} = 5.4 / 6.4 \times 100 = 84.4 \text{ ml}$$

$$\text{Salt} = 1 / 6.4 \times 100 = 15.6 \text{ ml}$$

3.3. Determination of pk_a

Principle

pk_a is the pH at which an acid is 50% ionized or dissociated.

Glassware and Equipments

pH meter, pipettes, burette, conical flask, beakers

Reagents

1. Acetic acid solution 1.4 g/l

Procedure

1. Titrate the unknown acid solution with the standard alkali and note the volume used V_1.

2. Take half amount of V_1.
3. Add it to another portion of acid and measure the pH.
4. The resulting pH is pk_a of acid solution.

From the table provided below, identify the acid as accurate as possible.

Table 5. Acids - Their molecular weight and pk_a values

Acid	Molecular weight	pk_a
Acetic acid	60.05	4.8
Benzoic acid	122.12	4.2
Lactic acid	90.08	3.9

4 LIPIDS

Lipids are heterogeneous group of compounds, which are soluble in non–polar solvents such as ether, chloroform, hexane and benzene, but insoluble in polar solvents. They can be classified in different ways. Based on the product of hydrolysis they can be of three types:

1. **Simple lipids:** These are the esters of fatty acids with various alcohols. Neutral lipids or fats or triglycerides and waxes are examples.
2. **Complex lipids:** These are esters of fatty acids with alcohol and some other groups such as phosphoric acid and sugars.
3. **Derived lipids:** They include hydrolysis products of above two classes and other compounds like sterols, fatty aldehydes, ketones, essential oils and vitamins (fat soluble).

Triglycerides (TG)

Fats mainly exist as triglycerides. Glycerol is the basic molecule. All the three hydroxyl (–OH) groups of glycerol are esterified with three fatty acids. If all the fatty acids are of same type, it is called simple TG and if all the three fatty acids are of different type, it is called mixed TG.

Fatty Acids

Fatty acids consist of a hydrocarbon chain with a carboxylic group at one end. Most naturally occurring fatty acids have an even number of carbon atoms and cis double bonds. They differ from each other in length of carbon chain and number of double bonds. Fatty acids with no, one and more double bonds are called saturated (SFA), monounsaturated (MUFA) and polyunsaturated (PUFA) fatty acids, respectively. Based on the position of double bonds between the methyl end and the nearest double bond, PUFA are further of two types: -ω -6 and ω -3. Irrespective of chain length of a fatty acid, methyl end carbon is called omega carbon. So, 18:2($\Delta^{9,12}$) is $\omega\oint-6$ fatty acid while 18:3($\Delta^{9,12,15}$) is $\omega\oint-3$ fatty acid.

Table 6. Some fatty acids and their common names

Symbol	Common or Trivial name	Systematic name	Structure
Saturated fatty acids			
10:0	Capric acid	Decanoic acid	$CH_3(CH_2)_8COOH$
12:0	Lauric acid	Dodecanoic acid	$CH_3(CH_2)_{10}COOH$
14:0	Myristic acid	Tetradecanoic acid	$CH_3(CH_2)_{12}COOH$
16:0	Palmitic acid	Hexadecanoic acid	$CH_3(CH_2)_{14}COOH$
18:0	Stearic acid	Octadecanoic acid	$CH_3(CH_2)_{16}COOH$
20:0	Aracidic acid	Eicosanoic acid	$CH_3(CH_2)_{18}COOH$
22:0	Behenic acid	Docosanoic acid	$CH_3(CH_2)_{20}COOH$
24:0	Lignoceric acid	Tetracosanoic acid	$CH_3(CH_2)_{22}COOH$
Unsaturated fatty acids			
16:1	Palmitioleic acid	cis Δ^9 hexadecenoic acid	$CH_3(CH_2)_5\,CH{=}CH\,(CH_2)_7\,COOH$
18:1	Oleic acid	cis Δ^9 octadecenoic acid	$CH_3(CH_2)_7\,CH{=}CH\,(CH_2)_7\,COOH$
20:1	Eicosenoic acid	cis Δ^{11} eicosenoic acid	$CH_3(CH_2)_7\,CH{=}CH\,(CH_2)_9\,COOH$
22:1	Erucic acid	cis Δ^{13} docosenoic acid	$CH_3(CH_2)_7CH{=}CH\,(CH_2)_{11}\,COOH$
18:2	Linoleic acid	cis $\Delta^{9,12}$ octadecedinoic acid	$CH_3(CH_2)_4CH{=}CHCH_2CH{=}CH(CH_2)_7\,COOH$
18:3	α–Linolenic acid	cis $\Delta^{9,12,15}$ octadecetrienoic acid	$CH_3CH_2CH{=}CHCH_2CH{=}CHCH_2CH{=}CH\,(CH_2)_7\,COOH$
18:3	γ–Linolenic acid	cis $\Delta^{6,9,12}$ octadecetrienoic acid	$CH_3(CH_2)_4CH{=}CHCH_2CH{=}CHCH_2CH{=}CH\,(CH_2)_4\,COOH$
20:4	Arachidonic acid	cis $\Delta^{5,8,11,14}$ eicosatetraenoic acid	$CH_3(CH_2)_4CH{=}CH(CH_2)CH{=}CHCH_2CH{=}CHCH_2CH{=}CH\,(CH_2)_3\,COOH$
20:5	Eicosapentaenoic acid	cis $\Delta^{5,8,11,14,17}$ Eicosapentaenoic acid	$CH_3(CH_2)CH{=}CH(CH_2)CH{=}CH(CH_2)CH{=}CHCH_2CH{=}CHCH_2CH{=}CH\,(CH_2)_3\,COOH$

Qualitative Analysis

4.1. Solubility Test

Principle

Non–polar lipids are soluble in non–polar solvents such as benzene, petroleum ether, and hexane, whereas polar lipids are soluble in polar solvents such as chloroform, methanol and ethanol. Oil or fats contain both polar and non–polar lipids in different proportions.

Glassware and Equipments

Test tubes, test tubes stand, pipette

Reagents

1. Ether
2. Chloroform
3. Benzene
4. Ethanol

Procedure

Take five dry test tubes. Take 2 ml each of water, ether, chloroform, benzene and ethanol in separate tubes. To each tube, add one drop of mustard or any other oil. Shake all the tubes well. Oil dissolves in all the tubes except for water. In water oil forms small droplets showing it is insoluble in water and floats on surface.

4.2. Emulsification

Principle

When oil is shaken with water, it becomes finely divided and is dispersed in water. Emulsifying agents such as bile salts, soaps, mono and diglycerides complete this process. These agents lower the surface tension of the liquid. It is important for fat digestion.

Glassware and Equipments

Test tubes, test tubes stand, pipette

Reagents

1. Bile salt solution

Procedure

Take two dry test tubes. In one tube take 2 ml of water and in another tube take 2 ml of bile salt solution. Add 2 drops of mustard oil in each tube and shake vigorously for 1 minute. In water, oil is broken in small pieces and floats on the surface. In bile solution, oil is seen suspended in small droplets. This is permanent emulsification.

4.3. Saponification

Principle

Oil or fat, when boiled with an alkali, is hydrolyzed and fatty acids are liberated, which form salts with alkali. This process is called saponification. Saponification number is number of mg of KOH required to saponify 1 g of fat.

Glassware and Equipments

Test tubes, test tubes stand, pipette, bunsen burner

Reagents

1. Na_2CO_3
2. NaCl
3. Concentrated HCl
4. $CaCl_2$ solution

Procedure

Take 4 ml of Na_2CO_3 solution. Add two drops of mustard oil. Shake vigorously and boil. A clean soapy solution is formed. Cool and divide the contents in three tubes.

- In the first tube, add a few drops of concentrated HCl. Acid hydrolyzes the soap and fatty acids get separated and float up.
- In the second tube, add NaCl in sufficient amount, soap precipitates and floats on the surface. It is called salting out of soap.
- In third tube, add a few drops of $CaCl_2$ solution, insoluble precipitate of calcium soap is formed.

4.4. Unsaturation Test

Principle

All neutral fats contain glycerides of some unsaturated fatty acids. These fatty acids become saturated by taking up iodine. If the fat contains more unsaturated fatty acids, it will take up more iodine. Iodine number is the number of grams of iodine taken up by 100 g of fat.

Glassware and Equipments

Test tubes, test tubes stand, pipette

Reagents

1. I_2 solution (alcoholic solution of I_2 containing $HgCl_2$)
2. Chloroform

Procedure

Take 10 ml of chloroform. Add 10 drops of Huble's I_2 solution. Solution of chloroform becomes pink in colour due to free I_2. Divide this solution in four separate tubes. Add mustard oil, coconut oil, dalda and groundnut oil in respective tubes. Add oil drop–wise shaking the tube vigorously for 30 s after each drop until pink colour is discharged. Count the number of drops. More is the number of drops required to discharge pink colour less is the unsaturation.

Note

- Pink colour is discharged because I_2 is taken by unsaturated oil.

4.5. Test for Cholesterol and / or streols

4.5.1. Liebermann–Burchard Reaction

Principle

H_2SO_4, when added to cholesterol in the presence of acetic anhydride, gives a green chromophore.

Glassware and Equipments

Test tubes, test tubes stand, pipette, dropper

Reagents

1. Concentrated H_2SO_4
2. Acetic anhydride
3. Chloroform

Procedure

Take 2 ml chloroform in a tube. Dissolve a little cholesterol. Add 10 drops of acetic anhydride and 1–3 drops of concentrated H_2SO_4. Mix well. Red colour appears, which changes through blue to green.

4.5.2. Salkowski's Test

Take 2 ml chloroform in a tube. Dissolve a little cholesterol. Add 2 ml of concentrated H_2SO_4 and shake gently. Upper chloroform layer shows red colour and lower H_2SO_4 layer shows a yellow colour.

4.5.3. Formaldehyde – H_2SO_4 Test

Take 2 ml chloroform in a tube. Dissolve a little cholesterol. Add 2 ml of formaldehyde–H_2SO_4 (1 part 40% formaldehyde: 50 parts H_2SO_4). Cherry colour develops. Pour off the chloroform in another tube. Add 2–3 drops of acetic anhydride. Blue colour develops. Test is more delicate than Salkowski's test.

4.6. Glycerol Test

Principle

Glycerol, a sweet, heavy liquid, is the major component of triglycerides and some other lipids. It has specific gravity of 1.26 at 20ºC. When it is heated with dehydrating agent such as $KHSO_4$ or P_2O_5, aldehyde (acrolein) is formed, which has irritating smell.

Glassware and Equipment

Balance, Test tubes, test tubes stand, pipette, bunsen burner

Reagents

1. $KHSO_4$

Procedure

Take 1 g of $KHSO_4$ in two tubes. In one tube add 2–3 drops of oil and in another tube add pure glycerol. Heat the test tubes. Note the smell. If the smell is pungent, the test is positive for glycerol.

Quantitative Analysis

4.7. Determination of Iodine Value of Oils

The double bonds of unsaturated fatty acids in glycerides undergo addition reactions with halogens. The amount of halogen absorbed by fat or oil is directly proportional to the degree of unsaturation. Iodine number is the number of grams of iodine taken up by 100 g of fat.

Principle

A known amount of fat or oil is dissolved in carbon tetrachloride and treated with excess of Wij solution (ICl) and then unreacted ICl is titrated as I_2 with hyposolution using starch as indicator.

Glassware and Equipments

Balance, burette, burette stand, stoppered conical flask (iodine flask) measuring cylinder

Reagents

1. 0.1N sodium thiosulphate [(Hypo) $Na_2S_2O_3$. 5 H_2O] solution: Dissolve 2.48 g of $Na_2S_2O_3.5H_2O$ in distilled water and make its volume to 100 ml.
2. Carbon tetrachloride
3. 10 % Potassium iodide: Dissolve 10 g KI in distilled water and make its volume to 100 ml.
4. Starch solution: Dissolve 1 g of starch by mixing with 10 ml of boiling water and add 90 ml of saturated sodium chloride solution.
5. Wij solution: Dissolve separately 8.5 g of iodine and 7.8 g of iodine monochloride in glacial acetic acid. Mix well and make its volume to 1 l.

Procedure

1. Weigh 0.5 g of fat or oil in a stoppered conical flask (iodine flask).

2. Add 10 ml of carbon tetrachloride to it, and shake well to dissolve the fat or oil.
3. Add 25 ml of Wij solution and replace the glass stopper.
4. After proper mixing, keep the flask in dark for an hour.
5. After 1 h, add 10 ml of 10% potassium iodide solution.
6. Titrate the liberated iodine with standard (0.1N) hyposolution to light yellow colour. Later add a few drops of starch as indicator titrate until the blue colour disappears.
7. Run a blank without sample in similar way.

Calculations

Iodine value = [(B – S) x N x 100] / W

Where, B = Volume (ml) of standard hyposolution for the blank

S = Volume (ml) of standard hyposolution for the sample

N = Normality of standard sodium thiosulphate solution (.0127)

W = Weight (g) of the sample

4.8. Determination of Acid Number of Fats and Oils

Acid number may be defined as milligram of potassium hydroxide required to neutralize the free fatty acid present in 1 g of oil.

Principle

Natural oils contain free fatty acids. Amount of free acid present gives an indication of the age and quality of the fat. The oils, which contain high amount of free fatty acids, are susceptible to oxidation and rancidification. Such oils are unfit for human consumption. Rancidification occurs as a result of peroxide formation at double bond by atmospheric oxygen and hydrolysis by microorganisms, with the liberation of acid and aldehyde. The free fatty acid present in fat or oil can be titrated and quantified using 0.1N KOH.

Glassware and Equipments

Balance, burette, burette stand, stoppered conical flask, water bath, measuring cylinder

Reagents

1. 95% ethanol

2. 0.1% Phenolphthalein: Dissolve 0.1 g phenolphthalein in absolute alcohol and make its volume to 100 ml.
3. 0.1 N KOH: Dissolve 5.6 g of potassium hydroxide in distilled water and make its volume to 1 litre.

Procedure

1. Take 1 g of fat or oil in a conical flask.
2. Add 25 ml of 95% ethanol and two drops of 0.1% phenolphthalein.
3. Boil the contents.
4. Titrate the contents of flask with N/10 KOH till faint pink colour appears.
5. Run a blank without sample in a similar way.

Calculations

Volume (ml) of 0.1 N KOH used to neutralize free fatty acid

$$V = T - B$$

Where,

T = Volume (ml) of KOH solution used for the test sample

B = Volume (ml) of KOH solution used for the blank

Weight of KOH per litre = 5600 mg

Milligram of KOH used to neutralize W g of oil = 5600xV/1000

Acid number of oil = [5.6xV] /W

4.9. Determination of Peroxide Value of Fats and Oils

Principle

Oxidative rancidity of fat or oil is caused by the action of air and generates peroxides. Degree of peroxide formation and time taken for development differ for oils or fats. Peroxide radical generation is determined by titration against hyposolution using potassium iodide as indicator.

Glassware and Equipments

Balance, Burette, burette stand, stoppered conical flask, measuring cylinder

Reagents

1. Solvent mixture, Chloroform–acetic acid mixture (1:2)
2. 5% Potassium iodide (KI): Dissolve 5 g KI in distilled water and make its volume to 100 ml.
3. 1% Starch solution: Dissolve 1 g of starch mixed with 10 ml of boiling water and add 90 ml of saturated sodium chloride solution.
4. Hyposolution $Na_2SO_3.5H_2O$ (N/500 diluted from N/10)

Procedure

1. Take 1 g of oil in a glass stoppered flask.
2. Add 20 ml of solvent mixture and add 1 ml of freshly prepared KI.
3. Shake the contents thoroughly.
4. Add 25 ml water to the conical flask and shake.
5. Titrate the contents of conical flask against hyposolution to light yellow colour. Later add a few drops of starch as indicator titrate until the blue colour disappears.
6. Similarly, run a blank without sample.

Calculations

Peroxide value (milliequivalents peroxide/kg) = $\frac{(T - B) \times N \times 1000}{W}$

Where,

T= volume (ml) of hyposolution used for the test sample

B = volume (ml) of hyposolution used for the blank

N = normality of hyposolution.

W =Weight of sample (g)

4.10. Determination of Saponification Number of Fats and Oils

Saponification number is defined as the number of mg of KOH required to saponify 1 g of fat or oil. It gives the average molecular weight of fatty acids in fat or oil. Longer the chain length of fatty acid in fat, less is the amount of KOH required /g of fat or oil to be hydrolyzed.

Principle

Oils or fats are hydrolyzed by using alkali, giving free glycerol and soap. The amount of unutilized alkali is determined by titration with standard hydrochloric acid. The alkali utilized for saponification is calculated by subtracting utilized alkali from the total used.

Glassware and Equipments

Balance, burette, burette stand, erlenmeyer, flask, cylinder and water bath, measuring cylinder

Reagents

1. Alcoholic KOH (0.5 N): Dissolve 40 g KOH in distilled ethanol and make its volume to 1 litre.
2. Hydrochloric acid (0.5 N): Take 42.4 ml of concentrated HCl and dilute to 1 litre.
3. 0.1% Phenolphthalein: Dissolve 0.1 g phenolphthalein in absolute alcohol and make its volume to 100 ml.

Procedure

1. Weigh 2.0 g of sample into a 250 ml Erlenmeyer flask.
2. Add 25 ml of alkali solution into the flask and mix the contents.
3. Fit condenser over the flasks and reflux on water bath till saponification is completed as indicated by appearance of clear solution.
4. Cool the flasks at room temperature and titrate excess of (unutilized) alkali with 0.5 N hydrochloric acid using phenolphthalein as indicator to colourless end point.
5. Similarly, run a blank without sample.

Calculations

Saponification number = (56.1 x (B – S) x N) / W

Where,

B = Volume (ml) of 0.5 N hydrochloric acid required for the blank

S = Volume (ml) of 0.5 N hydrochloric acid required for the sample

N = Normality of standard hydrochloric acid

W = Weight (g) fat or oil

4.11. Extraction of Lipids from Oilseeds using Soxhlet Method

Soxhlet method is the conventional method used for determination of total oil content. It is accurate, but large amount of seeds is required.

Principle

The oils and fats are completely miscible with non–polar organic solvents such as ether, petroleum ether and hexane. This property of lipids is used to extract them from tissues. It involves repeated extraction of the material with the solvent in the Soxhlet apparatus. The solvent is then distilled off completely. Oil (%) is calculated gravimetrically.

Glassware and Equipments

Balance, pestle and mortar, flasks, thimble, flash evaporator, Soxhlet apparatus

Reagents

1. Petroleum ether / hexane

Procedure

1. Take about 2 g of dried seed sample.
2. Grind the seed sample in a pestle mortar and pour it in filter paper or thimble.
3. Place the filter paper or thimble having sample in extractor of the Soxhlet apparatus.
4. Place some cotton wool at the top of the thimble to evenly distribute the solvent during extraction.
5. Weigh the flask and add organic solvents, two and a half times the capacity of the flask.
6. Extract oil for at least a period of 6–8 h till the solvent in the extractor becomes colourless.
7. Switch off the heaters and allow the flasks to cool. Evaporate the solvent using flash evaporator.
8. Weigh the flask containing oil.

Calculations

Weight of the seed $= W$ g

Weight of the empty flask $= W_1$ g

Weight of flask and lipid $= W_2$ g

Weight of lipid = $(W_2 - W_1)$ g
Lipid content (%) in seed = $(W_2 - W_1) / W \times 100$

Note

- Weigh finely crushed material in a precise and accurate manner.
- Put off flame in the laboratory while extracting oil using organic solvents.
- Add glass beads in the flask to check bumping.

4.12. Extraction of Lipids from Oilseeds using Cold Extraction Method

Principle

The oils and fats are completely miscible with non–polar organic solvents such as ether, petroleum ether, and hexane. This property of lipids is used to extract them from tissues. Finely ground sample is placed with solvent in the ratio of 1:20 (1 g sample: 20 ml solvent) for 24 h so that complete extraction takes place. To remove the water soluble impurities which come along with lipids, washings are given with 0.9% saline solution.

Glassware and Equipments

Glass stoppered bottle, pipette, balance, flash rotary evaporator, separatory funnel pestle and mortar, sintered glass funnel G-3, beakers

Reagents

1. Chloroform: methanol - Mix chloroform and methanol in 2:1 ratio.
2. 0.9% saline - Dissolve 0.9 g NaCl in distilled water and make its volume to 100 ml.
3. Anhydrous sodium sulphate (Na_2SO_4)

Procedure

1. Weigh accurately 10 g of oil seed and grind it in a pestle and mortar.
2. Add a pinch of anhydrous sodium sulphate (Na_2SO_4).

3. Transfer the contents in glass–stoppered bottle and to it add 200 ml of chloroform: methanol mixture, keep it for 24 h with intermittent shaking.
4. After 24 h filter the material through sintered glass funnel G–3 into a container, also give 2–3 washings with the same solvent.
5. Note the volume and take the filtrate into a separatory funnel.
6. Add 0.9% saline solution which should be $1/5^{th}$ volume of the total filtrate.
7. Shake the mixture vigorously and keep it overnight (12–24 h) without stopper.
8. Next day when the two layers get separate, remove the lower layer. Give washings with 10 ml of chloroform twice and collect the lower layers and pool them in a pre–weighed beaker.
9. Evaporate the pooled solvent containing lipids on the flash/rotary evaporator at 50–60ºC.

Calculations

Weight of seed sample=W g

Weight of empty beaker= W_1 g

Weight of beaker and lipid = W_2 g

Weight of lipid = $W_2–W_1$ g

Lipid content (%) in seed = $(W_2–W_1) / W \times 100$

4.13. Extraction of Lipids from Oilseeds using Cold Percolation Method

Principle

Cold percolation method involves the use of non–polar solvents like petroleum ether and hexane, which percolates through a column containing finely–powdered seeds along with anhydrous sodium sulphate. The oil gets eluted by the solvent, which is collected in a conical flask.

Glassware and Equipments

Mini glass column, balance, pestle and mortar, flasks, thimble, flash/rotary evaporator

Reagents

1. Petroleum ether / hexane
2. Anhydrous sodium sulphate (Na_2SO_4)

Procedure

1. Plug the percolator by putting cotton/glass wool at the bottom of it.
2. Put a layer of approximately 2 g of anhydrous sodium sulphate over the plug.
3. Take 2–3 g of finely powdered seed sample and mix well with 2–3 g of anhydrous sodium sulphate.
4. Pack the powdered material slowly inside the percolator. Put cotton over the material and cover with another layer of anhydrous sodium sulphate.
5. Fill the percolator with the solvent.
6. Let the solvent drop down under gravity. The solvent while flowing down will extract the oil from seed sample and collect the solvent in a flask.
7. Add more solvent to fill the percolator and allow it to pass through the column. Repeat this process thrice.
8. Evaporate the solvent using flash rotary evaporator.
9. Weigh the flask containing oil.

Calculations

Weight of the seed	= W g
Weight of the empty flask	= W_1 g
Weight of flask and lipid	= W_2 g
Weight of lipid	= $(W_2–W_1)$ g
Lipid content (%) in seed	= $(W_2–W_1) / W \times 100$

4.14. Nuclear Magnetic Resonance (NMR) spectroscopy of oil estimation

Oil estimation by NMR spectroscopy is a rapid, accurate and economical treated. Whole seed sample is analyzed allowing repeat measurements while maintaining the viability of seed for breeding purposes.

Principle

The NMR analyzer works on the principle of nuclear magnetic resonance spectroscopy. It determines the concentration of mobile hydrogen (protons) within the oil contained in the seed. When a sample is placed in the permanent magnetic field, the concentration of the protons is related to the degree of absorbance of radio frequency energy. However, a contribution is also made to the NMR signal by mobile hydrogen in moisture and bound hydrogen in the seed husk and seed oil. An electronic circuit is set up to discriminate between liquid and solid resonance. Moisture is adjusted by drying the seeds prior to analysis. The operations of NMR analyzer are controlled by a microprocessor.

Glassware and Equipments

NMR analyzer (Fig. 8), sample vials/holder

Reagents

1. Pure oil
2. Seed meal
3. Seed

Procedure

1. Insert sample holder filled with oil sample up to mark and calibrate the instrument, similarly insert sample holder filled with seed meal and calibrate the instrument.
2. Place the seed sample in sample holder and follow the instructions appearing on the display screen and enter the information via keypad.
3. After completion of the process, oil concentration (%) is displayed.
4. It takes 2–3 minutes for analyzing oil content of a sample.
5 The accuracy of oil estimation by NMR analyzer depends on magnet, sample temperature and preparation.
6. The application of NMR eliminates need for use of flammable solvents and minimizes random errors, although the estimates are often high.

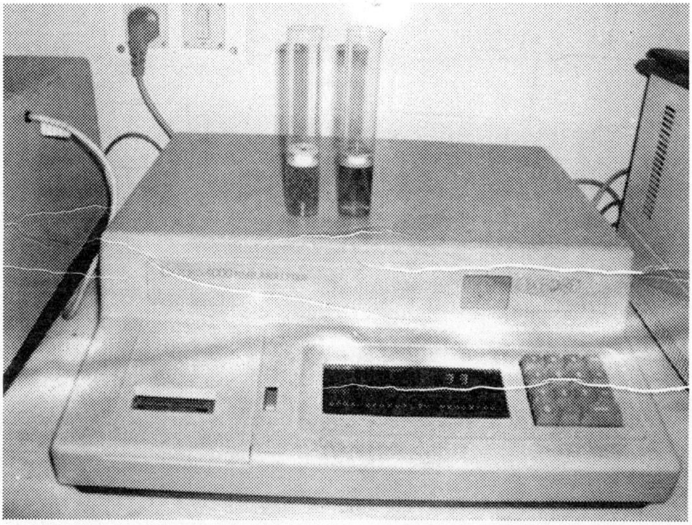

Fig. 8. NMR (Nuclear magnetic resonance) analyzer

4.15. Near Infrared Reflectance (NIR) Spectroscopy Method of oil estimation

The earliest applications of NIR spectroscopy were reported in the 1950s, but it was first adopted as an analytical technique by the agricultural food sector in 1974, when the Canadian Grain Commission replaced the traditional Kjeldahl method for determining protein in wheat with a new, NIR–based method. Since then this technique has been used for the industrial determination of protein, oil content in whole seed samples. Further, using this technique, oil content can be determined with smaller seed sample and more precisely than NMR analyzer.

Principle

The Infrared radiation is heat radiation. The energy of this radiation corresponds to energy differences between different vibrational modes in molecules. The near infrared spectroscopy is based upon the interaction of near infrared radiation (part of the electromagnetic spectrum that is between the visible and mid–infrared regions; 700–2500 nm) with matter. The reflected radiation from the sample surface at a given wavelength is calculated by measurement of reflectance of a standard reference at same wavelength. With the help of computer soft ware a calibration equation is developed which is used to determine the concentration of oil in the given sample.

Glassware and Equipments

NIR analyzer (Fig. 9), sample vials/cup

Reagents

1. Standard samples

Procedure

1. A very small size of sample is placed in sample cup.
2. It is scanned using NIR Spectrophotometer.
3. Using already prepared calibration oil content is quantified by the standard software.

Note

- Different types of NIR instruments are available in market for measurement of different parameters of given sample. Some models come pre-calibrated and in some models calibration has to be developed.

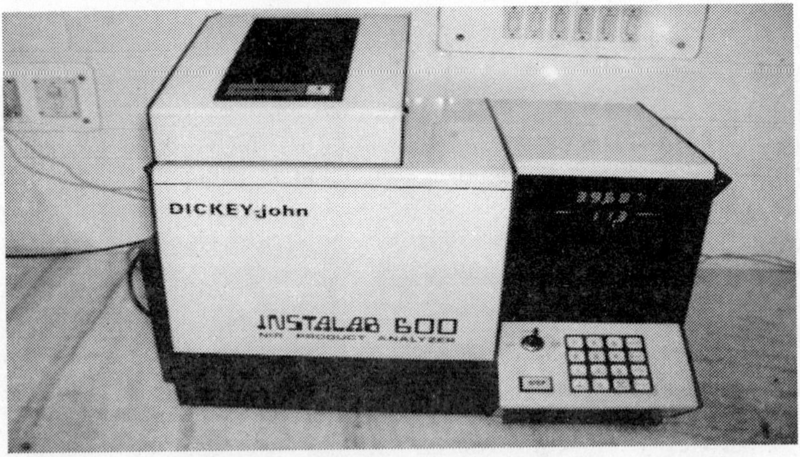

Fig. 9. NIR (Near infra red) analyzer

4.16. Determination of Unsaponifiable Matter

Principle

Sterols, higher aliphatic alcohol, pigments, vitamins and hydrocarbons are the major unsaponifiable matter as they are not saponified by caustic alkali but are soluble in oil/fat solvents. Saponified mixture is partitioned with water and hexane. The residue is dissolved in ethanol and titrated against sodium hydroxide.

Glassware and Equipments

Burette, burette stand, stoppered conical flask, water bath, Flat bottom flask with ground glass joint (250 ml), air condenser, separatory funnel, measuring cylinder, balance

Reagents

1. Alcoholic potassium hydroxide solution - Dissolve 40 g KOH in distilled ethanol and make its volume to 1 litre with 95% ethanol.
2. 0.1% Phenolphthalein - Dissolve 0.1 g of phenolphthalein in ethanol and make final volume to 100 ml.
3. Petroleum ether (60–80°C)
4. 10% ethanol - Mix 10 ml of ethanol in 90 ml of distilled water.
5. Sodium hydroxide solution (0.02 N) - Dissolve 0.8 g of NaOH in distilled water and make its volume to 1 litre.
6. Acetone
7. Anhydrous sodium sulphate

Procedure

1. Take 2 g of fat/oil sample in a 250 ml conical flask.
2. Add 25 ml of alcoholic potassium hydroxide solution.
3. Fit the condenser and reflux for about one hour.
4. Wash the condenser twice with water after saponification is complete.
5. Transfer the saponified mixture to a separatory funnel while it is warm.
6. Rinse the flask with ethanol and cold water. Transfer the contents to saponifed mixture.
7. Add 50 ml of petroleum ether and mix the contents. Two layers get separated.
8. Transfer the lower saponified layer to another separatory funnel.
9. Repeat the extraction thrice with petroleum ether.

10. Wash the combined petroleum ether extract three times with 25 ml of 10 % ethanol.
11. Wash the contents with 25 ml of distilled water and ensure that it is free from alkali.
12. Dry it over anhydrous sodium sulphate.
13. Evaporate the ether extract on vacuum evaporator to a constant weight.
14. Dissolve the contents in 50 ml of hot ethanol. Titrate with sodium hydroxide (0.02 N) using phenolphthalein as indicator.

Calculations

Weight (g) of free fatty acid in the extract expressed as oleic acid

$$= 0.282 \times V \times N$$

V = volume of standard sodium hydroxide solution

N = normality of sodium hydroxide solution.

$$\text{Unsaponifiable matter (\% by weight)} = 100\,[(A-B)/W]$$

Where,

A = Weight (g) of residue

B = Weight of free fatty acid in the extract

W = weight (g) of sample

4.17. Determination of Phospholipids from the Lipid Sample

Principle

Organically bound phosphorus is released as orthophosphate which can be extracted with 0.5 N HCl at 90°C. The liberated orthophosphate reacts with a mixed reagent of ascorbic acid and ammonium molybdate to form phosphomolybdic acid. This acid is then reduced by ascorbic acid to molybdenum blue complex. The absorbance of coloured complex is read at 820 nm. Amount of phospholipids is calculated by multiplying lipid phosphorus value with a factor of 25.

Glassware and Equipments

Balance, test tubes, test tube stand, water bath, air condenser, bunsen burner, oven, cyclo mixer, spectrophotometer pipette

Reagents

1. Reagent A (0.42% Ammonium molybdate) - Dissolve 0.42 g in 1 N H_2SO_4 and make its volume to 100 ml.
2. Reagent B (10% Ascorbic acid) - Dissolve 10 g ascorbic acid in distilled water and make its volume to 100 ml.
3. Reagent C - Mix A and B in the ratio of 6:1.
4. 10% Magnesium nitrate - Dissolve 10 g magnesium nitrate in 95% alcohol.
5. 0.5N HCl - Measure 42.4 ml of concentrated HCl (11.8 N) and dilute to 1000 ml with distilled water.

Procedure

1. Take 1 ml of lipid sample dissolved in suitable solvent.
2. Evaporate the solvent and dry the samples on a water bath.
3. Add 0.18 ml of 10% magnesium nitrate and shake the tubes.
4. Dry the sample at 100°C in an oven.
5. Ash the dried samples over strong flame till brown fumes ceases to come out of the tubes and the contents of tube become whitish.
6. Cool the tubes and add 1.8 ml of 0.5 N HCl.
7. Mix the contents on cyclomixer and reflux the samples in boiling water bath for 15 min.
8. Add 4.2 ml of reagent C.
9. Incubate the tubes at 45°C for 30 minutes.
10. Read the absorbance of blue colour developed at 820 nm.

4.18. Determination of Total Free Fatty Acids from Fats and Oils

Principle

Free fatty acids react with cupric acetate to form a bluish green–coloured complex. The absorbance of developed colour is read at 715 nm

Glassware and Equipments

Test tubes, test tubes stand, pipette, water bath, centrifuge, centrifuge tubes, vortex mixer, spectrophotometer

Reagents

1. Benzene
2. Cupric acetate (5%) - Dissolve 5 g of cupric acetate in distilled water and make final volume to 100 ml. Filter the contents and after filtration, set its pH to 6.0–6.2 using pyridine.

Procedure

1. Take 1 ml of free fatty acid containing lipid sample.
2. Evaporate the solvent by keeping the tubes in water bath at 60°C.
3. Add 5 ml of benzene and mix the contents on a cyclomixer vortex mixer.
4. Add 1 ml of 5% cupric acetate solution and mix the contents again on cyclomixer.
5. Centrifuge the contents at 3000 rpm for 5 minutes.
6. Separate the upper layer (blue in colour) using micropipette.
7. Read absorbance at 715 nm.
8. Prepare a standard curve using free fatty acid (palmitic acid) in the range of 2–10 μg.

4.19. Determination of Cholesterol / Sterols

Principle

Cholesterol reacts with acetic anhydride and concentrated sulfuric acid, to give a blue–coloured complex. However, the exact nature of the chromophore is not known. The absorbance of colour developed is read at 560 nm.

Glassware and Equipments

Test tubes, test tubes stand, pipette, water bath, spectrophotometer

Reagents

1. Chloroform
2. Developing reagent - Stock solution 5% $FeCl_3$ in acetic acid: Dissolve 5 g $FeCl_3$ in acetic acid to make volume to 100 ml. Working solution 0.05% $FeCl_3$ is prepared by taking 1 ml of stock solution and diluted to 100 ml using acetic acid.

3. Standard chlolesterol solution (200 μg/ml) - Dissolve 20 mg cholesterol in chloroform to make volume to 100 ml.

Procedure

1. Take 1 ml of lipid (0.5–1.0 mg) sample dissolved in chloroform.
2. Evaporate the solvent on a water bath.
3. Add 2 ml of colour developing reagent.
4. Keep the tubes at room temperature for 15 minutes.
5. Read the absorbance at 625 nm.
6. Prepare a standard curve using cholesterol in the range of 50–200 μg.

4.20. Determination of Glycolipids

Principle

Glycolipids contain one or more sugar residues attached to diacyl-glycerol or a sphingosine. Concentrated H_2SO_4 reacts with glycolipids to hydrolyze the glycosidic bond and release the sugars. These sugars can be estimated using phenol sulphuric acid method for estimation of sugars.

Glassware and Equipments

Test tubes, test tube stand, pipettes, measuring cylinder, beakers, volumetric flasks, spectrophotometer, water bath, distille unit, vortex mixer

Reagents

1. 5% distilled Phenol - Dissolve 5 g distilled phenol in distilled water to make volume to 100 ml.
2. Concentrated H_2SO_4
3. Standard glucose solution - Dissolve 100 mg glucose in distilled water and make the volume to 100 ml. This will serve as stock standard solution. From the stock take 10 ml and make the volume to 100 ml with distilled water to get working solution of 100μg/ml.

Procedure

1. Take 1 ml of lipid (0.5–1.0 mg) sample dissolved in chloroform.
2. Evaporate the solvent on a water bath.
3. Add 1 ml of distilled water and then add 1 ml of distilled phenol.
4. Mix the contents on a vortex mixter.
5. Add 4 ml of concentrated H_2SO_4.
6. Read absorbance at 490 nm.
7. Prepare a standard curve of glucose in the range of 10–100 μg.
8. Glycolipids are calculated from sugar concentration by multiplying with a factor of 4.45.

5 NUCLEIC ACIDS

The nucleic acids are the building blocks of living organisms. They are polymers consisting of monomers known as nucleotides. Each nucleotide contains three units, a sugar, a nitrogenous base, and a phosphate. Sugar and nitrogenous base joins to form nucleoside. The bond between the sugar and the base is called the glycosidic bond. Nucleic acids are divided into two classes on the basis of the sugar used to form the nucleotides. Ribonucleic acid (RNA) is built on a β-D-ribose sugar. Deoxyribonucleic acid (DNA) contains a deoxyribose sugar in which instead of the -OH group on the second carbon atom there is only H. The amines that form nucleic acids fall into two categories; purines and pyrimidines. There are three pyrimidines; cytosine (C), thymine (T), and uracil (U) and two purines; adenine (A) and guanine (G). DNA and RNA each contain four nucleotides. Both contain adenine and guanine and cytosine, but the fourth nucleotide in DNA is thymine, whereas RNA contains uracil. The carbon atoms in the sugar at the center of a nucleotide are numbered from 1′ to 5′. The -OH group on the 3′ carbon of one nucleotide can react with the phosphate group attached to the 5′ carbon of another nucleotide to form a dinucleotide held together by phosphate ester bonds. As the chain continues to grow, we get a polynucleotide.

5.1. Extraction of DNA

Principle

Cetyl trimethyl ammonium bromide (CTAB) is a cationic detergent. It dissolves cell membrane, denatures proteins and forms an insoluble complex with nucleic acids. Proteins, carbohydrates and other contaminants come in supernatant. Proteins attached with nucleic acids can be precipitated using chloroform, isoamyl alcohol. DNA, RNA–CTAB complex is precipitated with isopropanol. DNA pellet is dissolved in alcohol, dried, and re–dissolved in $T_{10}E_1$ buffer. RNA and polysaccharides are removed by RNase and ammonium acetate treatment, respectively.

Glassware and Equipments

Cooling centrifuge, centrifuge tubes, micropipettes

Reagents

1. Liquid nitrogen
2. Chloroform: isoamyl alcohol (24:1)
3. Isopropanol
4. 70% ethanol
5. $T_{10}E_1$ buffer: Take 1 ml of 1 M Tris of pH 8.0, add 200 μl of 0.5 M EDTA and make the volume to 100 ml.
6. DNA extraction buffer (DEB) - Refer to the Annexure for details.

Procedure

1. Grind 3–5 g of fresh or 0.5 g of lyophilized or dried plant material to a fine powder either using liquid N_2 or absolute alcohol.
2. Transfer the contents into 10 ml of pre–warmed (at 60ºC) DEB in capped polypropylene tubes.
3. Suspend the clumps using spatula and incubate for 1 h at 60°C with occasional mixing by gentle swirling in water bath.
4. Remove from water bath and add 1 ml of chloroform : isoamyl alcohol.
5. Mix by inversion to ensure emulsification of phases.
6. Centrifuge the contents at 15,000 rpm for 10 min at room temperature.
7. Take aqueous layer and transfer it to another tube using micropipette.
8. Add two volumes of absolute alcohol or 0.6 volumes of isopropanol into aqueous layer.
9. Collect the amorphous precipitates of DNA by centrifugation at 10,000 rpm for 10 min at 4°C.
10. Discard the supernatant and save the pellet.
11. Wash DNA precipitates with 70% alcohol. Keep for 20 min with gentle agitation.
12. Collect the pellet by centrifugation at 10,000 rpm at 4°C for 10 min. Discard the supernatant.

13. Invert the tubes on a paper towel, and dry the pellet overnight.
14. Dissolve the pellet in known volume of $T_{10}E_1$ buffer by keeping it for overnight at 4°C without agitation.

Note

- Care must be taken throughout all steps of DNA isolation to avoid contamination by DNases.
- All glassware should be clean.
- Wear latex gloves during handling DNA.
- Use lyophilized or freshly harvested sample.
- β-mercoptoethanol is used to inhibit polyphenol oxidase.

5.2. Removal of RNA using RNase Treatment

Principle

CTAB precipitates both DNA and RNA, so RNA is a major contaminant. RNA can be removed from DNA preparation by treating it with RNase. RNase should be free from DNases otherwise DNA may be degraded. RNase treatment is given at the rate of 50–100 μg/ml of crude DNA preparation.

Glassware and Equipments

Cooling centrifuge, centrifuge tubes, micropipettes

Reagents

1. RNase (DNase free)

Procedure

1. Add 50 μg/ml RNase (DNase free) into crude DNA preparation.
2. Mix it thoroughly and gently, and incubate at 37°C for 1h.
3. Add 0.3–0.4 ml of chloroform : isoamyl alcohol (24:1)
4. Mix thoroughly till an emulsion is formed.
5. Centrifuge the contents at 4°C for 15 min.
6. RNase is precipitated by this reagent.
7. Take supernatant avoiding whitish layer at interface.
8. To the supernatant, add two volumes of absolute alcohol and centrifuge at 10,000 rpm for 5 min at 4°C.

9. Discard the supernatant and take the pellet.
10. Wash the pellet with 70% alcohol and dry it overnight.
11. Dissolve DNA in $T_{10}E_1$ buffer.

5.3. Quantification and Purity of DNA

Principle

DNA and RNA both show absorption (optical density) of ultraviolet light at 260 nm. Proteins, which cause contamination of DNA, show absorption at 280 nm. The ratio of absorbance (A) at 260 nm and 280 nm provides an idea of purity of DNA extracts. A ratio of absorbance at 260 nm and 280 nm between 1.8–2.0 shows DNA is pure but values less than 1.8 indicate presence of proteins.

Glassware and equipments

Spectrophotometer, micropipette, test tubes, cuvette

Reagents

$T_{10}E_1$ buffer: Take 1 ml of 1 M Tris of pH 8.0, add 200 μl of 0.5 M EDTA and make the volume to 100 ml.

Procedure

1. Set blank using 3 ml of T_{10} E_1 buffer in cuvette.
2 Take 3 ml of buffer in test tube and remove 30 μl of buffer.
3. Add same amount (30 μl) of dilute DNA extract.
4. Take absorbance at 260 nm and 280 nm.

Calculations

Concentration of DNA (μg/ml) = 50 x A_{260} x dilution factor

$$\text{Purity of DNA} = \frac{A_{260}}{A_{280}}$$

5.4. Determination of DNA Content by Diphenylamine Method

Principle

DNA contains the basic sugar unit 2–deoxyribose. Strong mineral acids convert 2–deoxyribose to ω-hydroxylevulinyl aldehyde, which reacts with diphenylamine to give blue–coloured complex. The intensity of blue colour is read at 600 nm. This reaction is highly specific for deoxyribose.

Glassware and Equipments

Balance, uv–visible spectrophotometer, test tubes, test tube stand, pipettes, water bath

Reagents

1. Diphenylamine reagent: Take 150 mg diphenylamine in 10 ml of glacial acetic acid in a capped polypropylene tube and mix by repeated inversion until it is completely dissolved. Add 150 μl of concentrated sulphuric acid and mix, and then add 50 μl of acetaldehyde solution and again mix thoroughly.
2. Standard DNA solution (500 μg/ml): Dissolve 50 mg of DNA in 100 ml alcohol.

Procedure

1. Take 1 ml of DNA sample.
2. Add 4 ml of diphenylamine reagent.
3. Cover the tubes and mix by inverting the tubes.
4. Keep the tubes in boiling water bath for 10 min.
5. Cool and read the absorbance at 600 nm.
6. Prepare a standard curve using DNA in the range of 100–500 μg.

5.5. Determination of RNA Content by Orcinol Method

Principle

RNA contains ribose, which is converted into furfural in the presence of hot acids. Furfural in the presence of ferric ions reacts with orcinol to give green–coloured compounds, which can be read at 660 nm. This reaction is not as specific as diphenyl reaction. The reagent can react with deoxyribose also. If the DNA concentration is known, the exact concentration of RNA can be found out.

Glassware and Equipments

Balance, uv–visible spectrophotometer, test tubes, test tube stand, pipettes, waterbath

Reagents

1. 6% orcinol in 6% ethanol: Dissolve 300 mg orcinol in 6% aqueous ethanol (6 ml ethanol and 94 ml water) and make volume to 5 ml.

2. 0.1% $FeCl_3.6H_2O$: Dissolve 100 mg of ferric chloride in concentrated HCl and make volume to 100 ml.
3. Orcinol reagent: Mix 3.6 ml of 6% orcinol solution in 100 ml of 0.1% ferric chloride solution.
4. Standard RNA solution (500 μg/ml): Dissolve 50 mg RNA in 100 ml of distilled water.

Procedure

1. Take 1 ml of RNA sample.
2. Add 3 ml of orcinol reagent.
3. Cover the tubes and mix by inverting the tubes.
4. Keep the tubes in boiling water bath for 20 min.
5. Cool and read the absorbance at 660 nm.
6. Prepare a standard curve using RNA in the range of 100–500 μg.

5.6. To Amplify the Isolated DNA Sample using PCR

Principle

It is an *in vitro* method of enzymatic synthesis of specific DNA sequences, developed by Kary Mullis in 1983. It is a very simple technique for characterizing, analyzing and synthesizing any specific piece of DNA or RNA virtually from any organism, plant, animal, virus or bacteria. It exploits the natural function of enzyme polymerase, present in all living things to copy genetic material or perform molecular photocopying.

Glassware and Equipments

Agarose gel electrophoresis system, 1.5 ml tubes and rack, 200 μl PCR tubes and rack, 1–5 μl, 200 μl, 1.0 ml micropipette and tips, coolant rack, thermocycler machine (Fig. 10), UPS, gloves, UV–transilluminator

Reagents

1. Taq DNA polymerase
2. PCR buffer
3. dNTP's
4. Primers (forward and reverse)
5. $MgCl_2$

6. Distilled water (autoclaved nuclease free)
7. Genomic DNA template
8. Standard markers for PCR products (100 bp ladder)

Procedure

1. Prepare a reaction mix as given in Table 7.
2. Pour all the ingredients in PCR tubes.
3. Set programming on PCR machine as per instructions provided with machine.
4. Run PCR machine.
5. Once the time is over, remove the tubes.
6. Add 5 μl of gel loading buffer to each PCR tube.
7. Carefully pipette out 15 μl of reaction mixture and load on to 1.5% agarose gel. On the same gel, in another well, load 10 μl of ready to use markers.
8. Run the samples at 100 volts for 1–2 h or till the tracking dye bromophenol blue reaches 3/4th of the length of the gel.
9. Visualize the gel under UV transilluminator. If bands appear, take photograph of the gel and analyze it using Gel Documentation System and Software.

Table 7. Preparation of reaction mixture

S.N.	Component	Volume (μl)	Final concentration
1.	Autoclaved nuclease free ultra–filtered H_2O (pH 7.0)	16.0	-
2.	10 X PCR buffer	2.5	1X
3.	dNTP's (2 mM)	2.5	200–400 μM
4.	Primer F (20 μM)	0.5	0.4–0.8 μM
5.	Primer R (20 μM)	0.5	0.4–0.8 μM
6.	$MgCl_2$ (25 mM)	1.0	1–2.5 mM
7.	Taq DNA polymerase (1 unit/ μl)	1.0	0.8–1.0 unit
8.	DNA template (10μg)	1.0	0.2μg

Note

- While preparing reaction mixture, add DNA at last.
- Always wear gloves while handling gel.

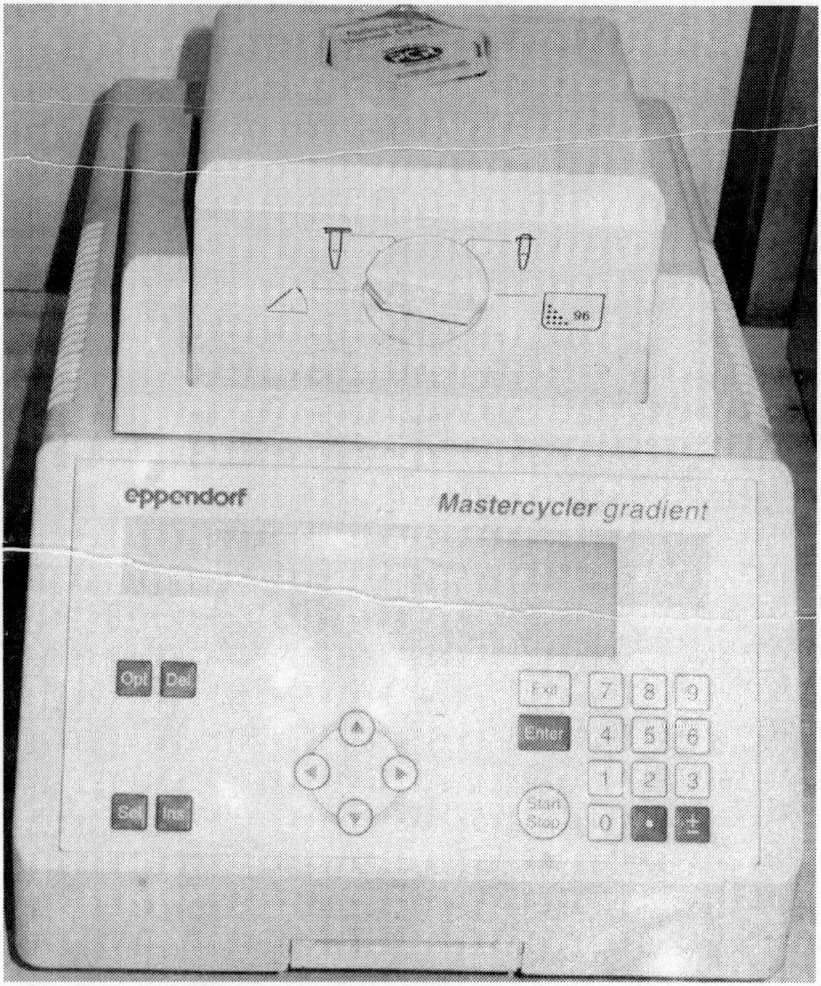

Fig. 10. Thermocycler

6 VITAMINS

Vitamins are organic compounds required by the body in trace amounts to perform specific cellular functions. They can be classified according to their solubility and functions in metabolism. The requirement of any given vitamin depends on organism. Not all vitamins are required by all organisms. Vitamins are not synthesized by humans and therefore must be supplied in the diet. Vitamins may be water soluble or fat soluble. Vitamin B (thiamine, riboflavin, biotin, pantothenic acid, folic acid, cobalamin, pyridoxine) and vitamin C (ascorbic acid) are water soluble, whereas four vitamins (Vitamins A, D, E and K) are fat-soluble.

Ascorbic acid is a strong reducing agent. It is widely distributed in animals and plants. Vertebrates such as humans and other primates, certain birds, and some fishes are unable to synthesize it. Vitamin C act as a coenzyme in hydroxylation reactions, it acts as an important antioxidant. A deficiency of ascorbic acid results in scurvy, a disease characterized by sore, spongy gums.

The E vitamins consist of 8 naturally occurring tocopherols, of which alpha–tocopherol is the most common. The primary function of vitamin E is as an antioxidant in prevention of the non–enzymatic oxidation of cell components by molecular oxygen and free radicals.

6.1. Determination of Vitamin C by Titration Method

Principle

Vitamin C (ascorbic acid) reduces 2, 6–dichloroindophenol dye to a colourless or leuco form. The ascorbic acid gets oxidized to dehydroascorbic acid and on titration with dye at the end point, pink colour appears.

Glassware and Equipments

Balance, burette, burette stand, pipette, volumetric flasks, dropper, centrifuge, centrifuge tubes, conical flasks.

Reagents

1. 4% Oxalic acid: Dissolve 4 g of oxalic acid in distilled water and make the volume to 100 ml.
2. 2, 6–dichloro indophenol: Dissolve 52 mg dye and 42 mg sodium bicarbonate in distilled water to make 200 ml and filter.
3. Standard ascorbic acid solution (0.2 mg/ml): Dissolve 20 mg ascorbic acid in 100 ml of 4% oxalic acid solution.

Procedure

1. Pipette 5 ml of the standard ascorbic solution in a conical flask.
2. Titrate it against dye till pink colour is obtained and remains for some time.
3. Note the volume of dye used for standard (V_1).
4. Take 0.5 g of sample and extract it twice with 4% oxalic acid.
5. Centrifuge the contents at 5,000 rpm, decant the supernatant and make the volume to 100 ml.
6. Take 5 ml of extract.
7. Titrate it against the dye solution till pink colour appears and persists.
8. Note the volume of dye used for sample (V_2).

Calculations

Strength of standard = 0.2 mg/ml

5 ml of standard = 1 mg of vitamin C

Volume of dye (V_1) used for standard = 1 mg of vitamin C

$$\text{Volume of dye } (V_2) \text{ for sample} = \frac{1\text{ mg} \times V_2}{V_1}$$

$$\text{Amount of Vitamin C (mg/100 g)} = \frac{1.0\text{ mg} \times V_2 \times 100\text{ ml} \times 100\text{ g}}{V_1 \times 5\text{ ml} \times W}$$

W = Weight of sample

6.2. Determination of Vitamin C by Colorimetric Method

Principle

Vitamin C (L-ascorbic acid) gets oxidized to its dehydroascorbic acid form by air especially at alkaline pH. However, it is stable in an acidic solution. Therefore, vitamin C is extracted in meta–phosphoric acid or in a mixture of meta–phosphoric acid and dilute acetic acid. Oxidized form of 2, 6-dichloroindophenol dye is blue in alkaline medium and red in an acidic medium. Reduced form of dye is colourless on the other hand and is termed as its leuco form. Pink colour developed in the reaction of ascorbic acid with dye in presence of N–amyl alcohol is read at 546 nm.

Glassware and Equipments

Balance, test tubes, test tube stand, pipettes, measuring cylinder, beakers, volumetric flasks, water bath, spectrophotometer, centrifuge, centrifuge tubes, mortar and pestle

Reagents

1. 2% meta–phosphoric acid - Dissolve 2 g of meta–phosphoric acid in distilled water and make volume to 100 ml.
2. 5% meta–phosphoric acid - Dissolve 5 g of meta–phosphoric acid in distilled water and make volume to 100 ml.
3. N–amyl alcohol
4. 2, 6–dichloroindophenol dye - Dissolve 52 mg dye and 42 mg sodium bicarbonate in distilled water to make 200 ml and filter.

Procedure

1. Take 1 g of fresh plant sample, homogenize in a mortar and pestle with 5 ml of 2% meta–phosphoric acid.
2. Allow it to stand for an hour.
3. After one hour, centrifuge the contents for 15 minutes at 2500 rpm.
4. Take the supernatant, repeat extraction with 5 ml of 2% meta–phosphoric acid.
5. Pool the supernatant and make volume to 10 ml with 2% meta–phosphoric acid.

6. Take 2 ml of supernatant and mix it with 4 ml of 5% meta–phosphoric acid.
7. Keep the tubes at room temperature without stirring for half an hour.
8. To this add 10 ml of N-amyl alcohol followed by 6.4 ml of 2,6-dichloro indole phenol dye.
9. Mix the layers by air bubbling through the lower layer.
10. Mix the layers by shaking vigorously.
11. Take out the upper layer for estimation of ascorbic acid.
12. Read absorbance at 546 nm against blank using 6.4 ml of distilled water instead of dye.

Calculations

The amount of ascorbic acid present in 1 ml of the original extract was obtained by using the regression formula:

$$Y = 0.1103-(0.14 \text{ X O.D.})$$

Where Y = concentration of Ascorbic acid in mg/ml of extract.

O.D. = Optical density

From the contents of 1 ml of the extract the ascorbic acid content per 100 g fresh weight could be calculated as follows:

$$\text{Free ascorbic acid} = \frac{(A \times V) \times 1000 \times 100}{W}$$

Where,

A = Y=mg Ascorbic acid /1 ml of original extract

V = Total volume of original extract

W = Weight of the plant sample (mg) used for analysis

6.3. Estimation of Total Tocopherols (Vitamin E)

Principle

Vitamin–E reduces ferric ions to ferrous ions which combine with bathophenanthroline to form an orange–coloured complex, which is read at 536 nm.

Glassware and Equipments

Balance, test tubes, test tubes stand, pipette, centrifuge, centrifuge tubes, vortex mixer, spectrophotometer, mortar and pestle

Reagents

1. Absolute ethanol
2. Purified xylene
3. 0.4% Bathophenanthroline reagent in absolute ethanol: Dissolve 100 mg bathophenanthroline in absolute ethanol and make volume to 25 ml.
4. Ferric chloride reagent: Dissolve 60 mg $FeCl_3.6H_2O$ in absolute ethanol and make volume to100 ml.
5. 85% Orthophosphoric acid: Dissolve 85 g orthophosphoric acid in absolute ethanol and make volume to100 ml.
6. dl α–tocopherol standard solution (0.02 mg/ ml of absolute ethanol)

Procedure

1. Crush 100 mg of sample in a pestle mortar.
2. Add 3 ml of ethanol.
3. Centrifuge the contents at 4,000 rpm for 10 min.
4. Pipette out 1.2 ml of supernatant, standard vitamin–E, and absolute alcohol in a glass stoppered centrifuge tubes marked as test, standard and blank, respectively.
5. Add 1.2 ml of distilled water to all the tubes and vortex thoroughly.
6. Add 1.2 ml of purified xylene and again vortex it for 2 min.
7. Centrifuge the tubes at 5,000 rpm for 5 min.
8. From each tube carefully pipette out 0.8 ml of xylene layer into a test tube containing 0.4 ml of bathophenanthroline reagent and mix thoroughly.
9. Add 0.4 ml of ferric chloride reagent followed by 0.4 ml of orthophosphoric acid.
10. Mix the contents in the tubes thoroughly after adding each reagent.
11. Read absorbance of test, standard and blank against xylene as reference at 536 nm immediately.

Note

- Avoid the samples to come in contact with sunlight.

- Do all the measurements quickly.

Calculations

$$\text{Total tocopherol (mg/100 g oil)} = \frac{A_T \times \text{conc. of standard} \times \text{total volume of sample (3 ml)} \times \text{percent}}{A_S \times \text{volume of test taken} \times \text{Wt. of tissue (g)}}$$

A_T = Absorbance of test– absorbance of blank

A_S = Absorbance of standard–absorbance of blank.

7 CHROMATOGRAPHY

Chromatography (colour and to write) is a physical method for separation of compounds. Tswett, Russian botanist has the credit for the development of chromatography. He first devised method to separate plant pigments using a tube filled with $CaCO_3$. Chromatography is based on the fact that sample distributes or partitions itself to different extents in two different, immiscible phases, which is referred as partition coefficient, K_d. If there are two immiscible phases A and B

K_d = Concentrationof sample in phase A / Concentration of sample in phase B

The two immiscible phases could be a solid and a liquid, or gas and a liquid, or a liquid and another liquid. One of the phases is a stationary phase (a solid, or a liquid supported on a solid) and does not move and the other is a mobile phase and move with respect to first. Mobile phase may be a liquid (liquid chromatography) or a gas (gas chromatography). All chromatographic methods involve passing a mobile phase through an immobile phase. Based on stationary phase there are two subgroups of gas chromatography and liquid chromatography, namely gas–solid chromatography and gas–liquid chromatography.

7.1. Thin Layer Chromatography

Principle

Separation of compounds is based on the competition of the solute and the mobile phase for binding on to the stationary phase. A mixture of compounds can be separated using Thin Layer Chromatography (TLC). The separation takes place due to differences in the adhesion force of the molecules that are present in the mixture to a mobile phase (normally a solvent) and to a stationary phase (thin layer, silica gel).

Glassware and Equipments

TLC plates, TLC chambers, uvlamp, capillary tube, pipette, spraying flask, oven

Reagents

1. Solvent system
2. Ceric ammonium molybdate (CAM) solution: Dissolve 5 g cerium sulphate and 25 g ammonium molybdate in 450 ml of water and 50 ml of H_2SO_4. It is used as a spray reagent.
3. Silica gel G : Prepare a slurry by lissolving 30 g of silica gel G in 60 ml of distilled water.

Procedure

1. Prepare a TLC plate using silica gel G, dry it and activate it.
2. Dissolve the compound in appropriate solvent (0.1–10 ml/ml).
3. Spot a small amount of compound 0.5 cm from bottom of the plate. Allow the solvent to evaporate.
4. Take the solvent and pour it in the TLC chamber.
5. Allow the solvent to equilibrate inside the TLC chamber.
6. Place the TLC in the chamber and let the solvent front run up the plate.
7. When the plate has run far enough (solvent front 0.5–1cm from top of plate), remove the plate from the TLC chamber.
8. Let the plate dry and then visualize.
9. For visualization, check under UV lamp and mark any UV active spots.
10. Stain the plates by spraying or dipping them in a CAM solution.
11. Heat the plate on a hot plate or in an oven for 10 seconds to 2 minutes.
12. Compare Retention factor values of spots with Rf values of standard compounds.

$$Rf = \frac{\text{Distance travled by spot from origin (cm)}}{\text{Distance traveled by solvent from origin (cm)}}$$

Note

- Place a filter paper in the TLC chamber that reaches from top to bottom as this will absorb the solvent and prevent evaporation of the solvent.
- Always note any UV active spots.

- Mark the plate from backside with a marker.
- Now-a-days, ready–to–use plates are available in the market.
- Activate the protein by placing in oven at 110°C for 1 h.

7.2. Paper Chromatography

Principle

It is a partition chromatography. Cellulose is used as supporting medium over which the solvent flows. Water is stationary phase and organic solvents act as mobile phase. As the solvent passes through an area of the paper, the solute partitions itself between the aqueous and the organic phases, depending on its relative solubility in two phases. If it is soluble in water it will move slowly compared to the solvent front but if it has more affinity for organic solvents it will move fast along with them.

The distance moved by the solute in relation to the distance moved by the solvent is known as retention factor (Rf).

$$Rf = \frac{\text{Distance moved by the solute}}{\text{Distance moved by the solvent front}}$$

Glassware and Equipments

Whatman no. 1 paper strips, paper chromatography chamber, capillary tube/micropipette

Reagents

1. Butanol:Acetic Acid:water (6:2:2) solvent mixture for amino acids
2. 0.25% ninhydrin: Dissolve 250 mg ninhydrin in acetone and make the volume to 100 ml. It is used as a spray reagent.

Procedure

1. Prepare strips of 43 x 12 cm from the Whatman no. 1 filter paper sheet. Width of the strip can be changed according to the number of samples to be applied. Keep 1.5 cm for each sample.
2. Mark the sheet with lead pencil at the distances and folds as shown in Figure 11.

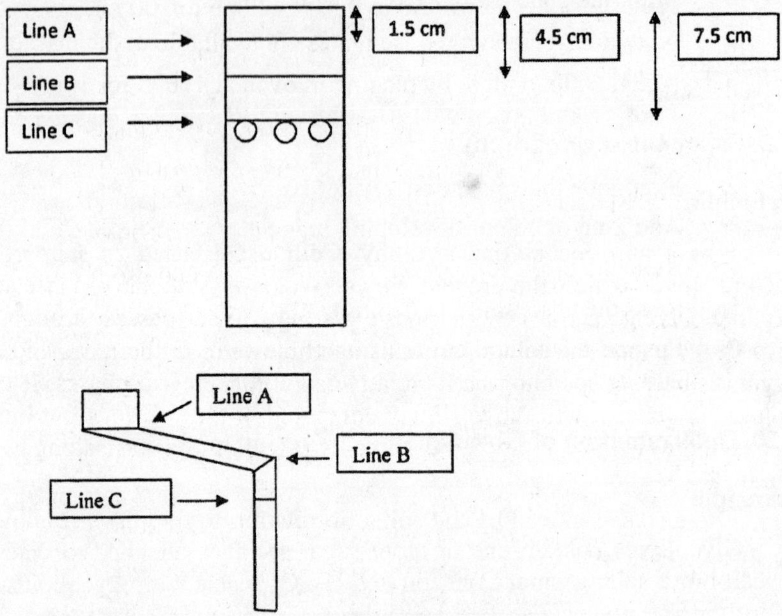

Fig. 11. Paper chromatography.

3. Dissolve the compound in 50% ethanol (0.1–10 ml/ml).
4. Spot a small amount of sample mixture along with standard mixture on line C with the help of capillary tube/micropipette.
5. Place the paper in paper chromatography chamber.
6. Develop the paper strip in ascending Chromatography method.
7. Allow the solvent to run.
8. After completion of the process, remove the paper and dry it at room temperature.
9. Use 0.25% ninhydrin solution in acetone as staining solution for mixture of amino acids.
10. Dry the chromatogram and heat at 60°C for 5–10 min.
11. Find Rf values of ninhydrin positive spots (bluish violet in colour) of standard as well as test samples.

Note

- Spot should not exceed in diameter from 0.5 cm.

- Choice of solvent generally depends on the type of mixture to be studied. To obtain good results, choice of solvent plays a key role.
- For Rf values of standard amino acids and sugars, refer to Table 1.3 from Annexure .
- In ascending technique, the solute and solvent move against gravity by capillary action. While in descending technique, the solute and solvent move in the direction of the gravity.
- Different types of paper can be used e.g., Pure cellulose (Whatman no. 1 and 3), Modified cellulose (DEAE cellulose, CM cellulose etc), glass fibre paper, Whatman GF/A.
- For separation of sugars, use solvent mixture Butanol: Acetic Acid: water (4: 1: 5), and spray the paper strip in reagents as per their order given below:
  - i. $AgNO_3$–acetone solution - Take 1 ml of saturated $AgNO_3$ solution and dilute it to 200 ml with acetone along with shaking using magnetic stirrer.
  - ii. 0.2 M alcoholic NaOH solution - Dissolve 2 g NaOH in small quantity of water and make the volume to 100 ml with ethanol.
  - iii. 20% sodium thiosulphate solution.

7.3. Gas Liquid Chromatography

Identification and quantification of fatty acids in fat or oil is done using gas liquid chromatography (GLC). This technique facilitates quantification of even the traces of fatty acid present in oil in seeds or cotyledons.

Principle

Fatty acids are made volatile by converting them into methyl esters directly or by trans–esterification. Esterification can also be done after saponification. The esters of fatty acids are identified and quantification is done by comparing with a set of standard fatty acid esters. A number of column like 10% FFAP chromosorb W(80–100 mesh)/6% BDS chromosorb W (100–120 mesh) / SP-2300 (2%) + SP-2310 (3%) are generally used with flame ionization detector. Nitrogen gas (flow rate 40-50 ml/minute) is commonly used as a carrier and hydrogen (flow rate 34–40 ml/ minute) and zero air (flow rate 275–300 ml/minute) are used for ignition of flame in detector. Injector and detector temperature depends on the nature of the column used and generally varies from 220–300°C. Generally, the injector and detector temperatures are kept 50°C above column temperature. Isothermal as well as temperature gradient methods can be used for the separation of fatty acids. However, temperature gradient provides better separation.

Glassware and Equipments

Screw capped vial, microlitre syringe, Gas chromatograph (GC)/ Gas chromatograph –mass spectrometer (GC-MS), balance, reagent bottles, mortar and pestle, vortex mixer

Reagents

1. Petroleum ether (40–60°C) / hexane
2. Reagent A - Sodium hydroxide in methanol: Dissolve 40 mg NaOH in 50 ml of methanol.
3. Reagent B - 8% Sodium Chloride: Dissolve 8 g NaCl in 100 ml of distilled water.

Procedure

Preparation of methyl esters

1. Crush 10–15 oven dried seeds in a mortar and pestle.
2. Transfer finely powdered sample to an air tight screw capped vial and add 0.5 ml petroleum ether/ hexane to the vial.
3. Keep the vials at room temperature for at least 1.5 hours.
4. Transfer the liquid in another air tight screw capped vial discarding the meal.
5. Add 0.5 ml of reagent A to each tube and vortex it.
6. Keep the tubes at room temperature for 45 minutes.
7. Add 0.75 ml of reagent B to each tube and vortex it.

Injection of methyl esters to GLC column

1. Add 1 μl of upper layer using micro liter syringe to GLC column.
2. Measure the retention time and identify individual fatty acids by comparing the profile with the retention time of standard methyl esters. Esters appear in the order of increasing number

of carbon atoms and of increasing unsaturation for the same number of carbon atoms (Fig. 12).

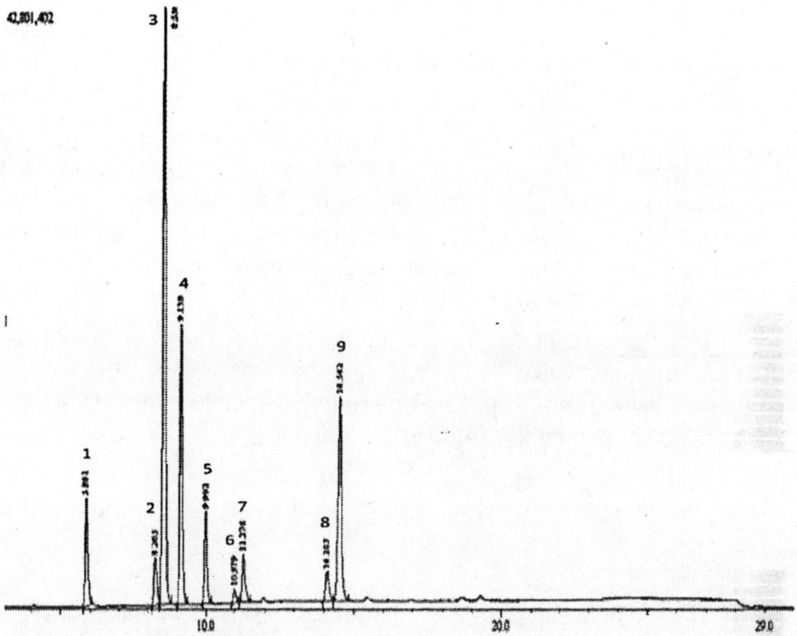

Fig. 12 Fatty acid profile of oil sample

8 ENZYMES

Enzymes

Enzymes are biological catalysts, required in small quantities, can transform chemical reactions at a very high rate. They are protein in nature with the exception of ribozymes, where RNA acts as enzyme. They are specific in action.

Most of the protein enzymes require cofactors for their activities. Protein part of enzyme is called apoenzyme. Non–protein part of enzyme is called prosthetic group. If prosthetic group is a metal ion, it is called as a cofactor and if prosthetic group is a vitamin then it is called as co–enzyme.

Enzyme along with prosthetic group is known as holoenzyme,

Enzyme classification

All the enzymes have been classified on the basis of reaction they catalyse. Each enzyme is represented by a name and EC (Enzyme commission) number of four digits. First digit represents the class, second corresponds to sub–class, third stands for sub–subclass or sub–group and fourth digit represents the number of particular enzyme in the list.

Enzymes have been grouped in six classes as given below:

1. Oxidoreductase
2. Transferase
3. Hydrolase
4. Lyase
5. Isomerase
6. Ligase

Factors affecting Enzyme reaction

Numbers of factors affect the enzyme reaction. Concentration of enzyme, concentration of substrate, pH of reaction mixture, temperature, products of reaction, time given for reaction affect the rate of reaction.

Light and other physical factors also affect the rate of an enzyme catalyzed reaction.

There are certain substances which inhibit the rate of a reaction and they are called as Inhibitors. Inhibitors can be of three types; competitive, non–competitive and uncompetitive.

Enzyme units

Enzyme activity is measured in terms of the enzyme activity unit (U). It is defined as the amount which will catalyse the transformation of 1 μmole of the substrate per minute under standard conditions.

Another unit is katal (kat) - It is defined as the amount which will catalyse the transformation of 1 μmole of the substrate per second.

1 kat = 6×10^7 U

1 kat = 10^9 nanokat

1 U = 16.67 nanokat

Activity

It refers to the total units of enzyme in the sample.

Specific activity

It is the number of enzyme units per millgram of protein. It is a measure of the purity of enzyme. During enzyme purification, specific activity increases and its value becomes maximal and constant when the enzyme is pure.

8.1. Estimation of Amylase Activity in the Given Sample

Principle

Amylases hydrolyse starch and glycogen into simple sugars. There are two types of amylases - α–amylase and β–amylase. α–amylase is an endohydrolase, which hydrolyses α -1, 4-linkage within the chain and breaks these polymers into small fragments. In contrast β–amylase is an exohydrolase and it acts from non–reducing end removing two glucose molecules at a time. The hydrolytic action of β–amylase results in increasing the reducing power of the starch solution due to the release of large number of molecules of the reducing sugar (maltose). The reducing sugars are reduced to red–brown coloured product by their reaction with DNSA reagent. The absorbance of developed colour can be determined at 540 nm.

Glassware and Equipments

Balance, water bath, spectrophotometer, test tubes, beakers, measuring cylinder, pipettes, test tube stand, centrifuge, centrifuge tubes, pestle and mortar

Reagents

1. Dinitrosalicylic acid (DNSA) reagent: Dissolve 10 g of 3,5-dinitrosalicylic acid, 2.0 g phenol and 0.5 g sodium sulphate in 1% NaOH solution and make the volume to 1 l.
2. Sodium potassium tartarate (40%): Dissolve 40 g of sodium potassium tartarate in distilled water and make the volume to 100 ml.
3. Starch (1%, w/v): Take 1 g of starch and add 100 ml of boiling water to it. Boil the solution for 2–3 min. Cool it, filter it and use as a substrate.
4. Sodium phosphate buffer (0.05 M, pH 7.0) containing 0.5 M NaCl. Refer Table 1.7 in annexure. Double dilute 0.01 M buffer to get 0.05 M buffter. Add 2.92 g NaCl in 100 ml of buffer.
5. Standard maltose solution (1 mg/ml): Dissolve 100 mg of maltose in 100 ml of distilled water.

Procedure

1. Take 500 mg of washed germinated barley seeds. Macerate the seeds in a chilled pestle and mortar by adding 5 ml of phosphate buffer (pH 7.0) containing 0.5 M NaCl.
2. Centrifuge the contents at 10,000 x g for 15–20 min in a cooling centrifuge at 4°C.
3. Decant the supernatant and use it as crude enzyme extract.
4. Take 1 ml of starch and 1 ml of dilute enzyme solution in two tubes. Label one tube as control and other as test sample.
5. Add 2 ml of DNSA reagent to the tube labelled as control.
6. Incubate the other tube at 37°C for 15 min and then stop the reaction by adding 2 ml of DNSA reagent.
7. Keep both the tubes in a boiling water bath for 5 min and then add 1 ml of sodium potassium tartrate to each of the tubes.
8. Cool the tubes under running tap water. Make the volume to 15 ml with distilled water.

9. Take absorbance at 540 nm using control as blank.
10. Prepare a standard curve of maltose in the range of 0.2–1.0 mg/ml.

8.2. Study the Effect of Temperature on Amylase Activity and Find Out Optimum Temperature for its Activity

Principle

All the enzymes have temperature range for their activity. To determine the enzyme activity in the reaction mixture, enzyme reaction should be carried out at optimum temperature, where enzyme and substrate binding is more feasible as compared to rest of the temperatures.

Glassware and Equipments

Balance, water bath, spectrophotometer, test tubes, beakers, measuring cylinder, pipettes, test tube stand, centrifuge, centrifuge tubes, pestle and mortar

Reagents

1. Dinitrosalicylic acid (DNSA) reagent: Dissolve 10 g of 3,5-dinitrosalicylic acid, 2.0 g phenol and 0.5 g sodium sulphate in 1% NaOH solution and make the volume to 1 l.
2. Sodium potassium tartarate (40%): Dissolve 40 g of sodium potassium tartarate in distilled water and make the volume to 100 ml.
3. Starch (1%, w/v): Take 1 g of starch and add 100 ml of boiling water to it. Boil the solution for 2–3 min. Cool it, filter it and use as a substrate.
4. Sodium phosphate buffer (0.05 M, pH 7.0) containing 0.5 M NaCl some as in previous exercise.
5. Standard maltose solution (1 mg/ml): Dissolve 100 mg of maltose in 100 ml of distilled water.

Procedure

1. Take 500 mg of washed germinated barley seeds. Macerate the seeds in a chilled pestle and mortar by adding 5 ml of phosphate buffer (pH 7.0) containing 0.5 M NaCl.
2. Centrifuge the contents at 10,000 x g for 15–20 min in a cooling centrifuge at 4°C.

3. Decant the supernatant and use it as crude enzyme extract.
4. Take 1 ml of starch and 1 ml of dilute enzyme solution in six tubes in a set of triplicate. Label one tube in each set as control and other as test samples.
5. Add 2 ml of DNSA reagent to the tube labelled as control.
6. Incubate the different reaction mixture tubes at different temperatures such as 20°C, 25°C, 30°C, 37°C and 45°C for 15 min.
7. Stop the reaction in each tube by adding 2 ml of DNSA reagent.
8. Keep all the tubes in a boiling water bath for 5 min and then add 1 ml of sodium potassium tartrate to each of the tubes.
9. Cool the tubes under running tap water. Make the volume to 15 ml with distilled water in each tube.
10. Take absorbance at 540 nm using control as blank.
11. Prepare a standard curve of maltose in the range of 0.2–1.0 mg/ml and calculate the enzyme activity.
12. Plot a graph of the enzyme activity with temperature and determine temperature optima.

8.3. Determination of Catalase (EC 1.11.1.6) Activity

Principle

Catalase is able to use one molecule of H_2O_2 as substrate or electron donor and another molecule of H_2O_2 as oxidant or e^- acceptor and catalyses following reaction:

$$2H_2O_2 \xrightarrow{CAT} 2H_2O + O_2$$

On decomposition of H_2O_2, its absorption decreases with time and decrease in absorbance at 240 nm per unit time is a measure of the catalase activity.

Glassware and Equipments

Test tubes, test tube stand, pipettes, measuring cylinder, beakers, volumetric flasks, uv–vis spectrophotometer, centrifuge, centrifuge tubes, pestle and mortar

Reagents

1. 0.05 M phosphate buffer (pH 7.5) - See annexure Table 1.7. Double dilute 0.1 M phosphate buffer before use to get 0.05 M buffer. Add 1 g polyvinyl pyrroliodone in 100 ml of buffer.
2. H_2O_2 solution (39 mM) - Dilute 0.2 ml of 30% H_2O_2 to 50 ml with phosphate buffer 0.05 M (pH 7.5).

Procedure

1. Homogenise100 mg of leaf sample in 2.5 ml of 0.05 M phosphate buffer (pH 7.5) containing 1% PVP (polyvinyl pyrrolidone).
2. Centrifuge the homogenate at 15000 x g for 30 minutes at 4°C.
3. Keep the supernatant for enzyme assay.
4. In the cuvette, take 1.9 ml of 0.05 M phosphate buffer (pH 7.5).
5. Add 0.1ml of enzyme extract.
6. Initiate the reaction by adding 1ml of H_2O_2.
7. Record the utilization of H_2O_2 at intervals of 30 seconds for 3 minutes by measuring the decrease in absorbance at 240 nm.
8. Express CAT activity as μmoles of H_2O_2 decomposed min^{-1} mg^{-1} protein or min^{-1} g^{-1} of fresh weight.

Note

- Extinction coefficient of H_2O_2 has the value of 0.0394 mM^{-1} cm^{-1}. This indicates one micromole of H_2O_2 per ml will have an absorbance of 0.039.
- For taking absorbance in UV range use quartz cuvette.

8.4. Determination of Ascorbate Peroxidase (EC 1.11.1.11) Activity

Principle

Ascorbate peroxidase catalyses the conversion of H_2O_2 to water at the expense of electron acceptor ascorbate, which gets converted into mondehydroascorbate (MDA).

$$2H_2O_2 + \text{ascorbate} \xrightarrow{\text{APX}} H_2O + \text{Monodehydroascorbate}$$

Ascorbic acid has absorption maxima at 265 nm. Oxidation of acid causes decrease in its absorption which is measured with time.

Glassware and Equipments

Balance, test tubes, test tube stand, pipettes, measuring cylinder, beakers, volumetric flasks, uv–vis spectrophotometer

Reagents

1. 0.05 M phosphate buffer (pH 7.5) - see annexure Table 1.7. Double dilute 0.1 M phosphate buffer before use to get 0.05 M buffer.
2. 0.5 mM ascorbic acid - Dissolve 8.8 mg ascorbic acid in distilled water and make final volume to 100 ml.
3. H_2O_2 solution (39 mM) - 0.2 ml of 30% H_2O_2 diluted to 50ml with phosphate buffer 0.05 M (pH 7.5).

Procedure

1. Take 1 ml of 0.05 M phosphate buffer (pH 7.5), 0.8 ml of 0.5 mM ascorbic acid, 0.2 ml of enzyme extract as prepared in 8.3 Ex. and 1 ml of H_2O_2 solution.
2. Record the decrease in absorbance at 30 seconds interval for 3 minutes at 265 nm.
3. Ascorbate peroxidase is expressed as nmoles of MDA formed min^{-1} mg^{-1} protein or min^{-1} g^{-1} of fresh weight of tissue.

Note

- Extinction coefficient of MDA has the value of 2.8 mM^{-1} cm^{-1}. This indicates one micromole of MDA per ml will have an absorbance of 2.8.
- For taking absorbance in UV range use quartz cuvette.

8.5. Determination of Glutathione reductase (EC 1.4.1.2) Activity

Principle

Glutathione reductase catalyzes the reduction of oxidised glutathione (GSSG) to reduced glutathione (GSH) in a NADPH dependent reaction at the expense of two protons and two electrons.

$$GSSG + NADPH + H^{+} \xrightarrow{GDH} 2GSH + NADP^{+}$$

Glassware and Equipments

Balance, test tubes, test tube stand, pipettes, measuring cylinder, beakers, volumetric flasks, uv–vis spectrophotometer, centrifuge, centrifuge tubes pestle and mortar

Reagents

1. 0.1 M phosphate buffer (pH 7.5) - See annexure Table 1.7.
2. 0.2 mM EDTA - Dissolve 7.44 mg ethylene diamine tetra acetic acid disodium salt (EDTA) in distilled water and make final volume to 100 ml.
3. 1.5 mM $MgCl_2$ - Dissolve 30.45 mg $MgCl_2$ in distilled water and make final volume to 100 ml.
4. 0.025 mM NADPH - Dissolve 2.08 mg NADPH in distilled water and make final volume to 100 ml.
5. 0.25 mM glutathione oxidised - Dissolve 15.31 mg of oxidised glutathione in distilled water and make final volume to 100 ml.

Procedure

1. Homogenise 200 mg of sample with 4 ml of 0.1 M phosphate buffer (pH 7.5) containing 1% PVP (polyvinyl pyrrolidone), 1 mM EDTA (ethylene diamine tetraacetic acid) and 10 mM β-mercaptoethanol using pre-chilled pestle and mortar.
2. Centrifuge the homogenate at 12000 x g for 15 minutes at 4°C.
3. Keep the supernatant for enzyme assay of glutathione reductase, superoxide dismutase.
4. Take 0.4 ml of 0.1 M phosphate buffer (pH 7.5), 0.2 ml of EDTA (0.2 mM), 0.2 ml of $MgCl_2$ (1.5 mM), 0.4 ml of NADPH (0.025 mM) and 0.8 ml of enzyme extract.
5. Start the reaction by addition of 0.4 ml of oxidized glutathione (0.25 mM).
6. Mix the contents gently.
7. Record the decrease in absorbance at 340 nm at interval of 30 seconds for 3 minutes.
8. Express the enzyme activity as n moles of NADP formed min^{-1} mg^{-1} protein.

Note

- The molar extinction coefficient for NADPH is 6.22 $mM^{-1}cm^{-1}$ This indicates one micromole of NADPH per ml will have an absorbance of 6.22..
- For taking absorbance in UV range use quartz cuvette.

8.6. Determination of Superoxide dismutase (EC 1.15.1.1) Activity

Principle

Superoxide Dismutase catalyses the disproportionation of superoxide anion ($O_2^{\cdot -}$) to H_2O_2 and molecular O_2.

$$2O_2^{\cdot -} + 2H^+ \rightarrow H_2O_2 + O_2$$

The assay of SOD is based on the ability of this enzyme to inhibit the auto–oxidation of pyrogallol in presence of EDTA.

Glassware and Equipments

Balance, test tubes, test tube stand, pipettes, measuring cylinder, beakers, volumetric flasks, spectrophotometer

Reagents

1. 0.1M Tris-HCl buffer (pH 8.2) - See annexure Table 1.8
2. 6 mM Pyrogallol solution - Prepare fresh by dissolving 75.66 mg pyrogallol in distilled water and make final volume to 100 ml.
3. 6 mM EDTA - Dissolve 223.3mg ethylene diamine tetra acetic acid disodium salt (EDTA) in distilled water and make final volume to 100 ml.

Procedure

1. To the cuvette add 1.5 ml of 0.1 M Tris HCl buffer (pH 8.2), 0.5 ml of 6 mM EDTA, 1 ml of 6 mM pyrogallol solution and 0.1 ml enzyme extract as prepared in 8.5 Ex.
2. Record the change in absorbance at 420 nm at 30 seconds interval for 3 minutes.
3. Prepare a blank consisting of reaction mixture without enzyme extract.

4. A unit of enzyme activity is defined as the amount of enzyme causing 50% inhibition of auto–oxidation of pyrogallol observed in blank.

8.7. Determination of Glutamate dehydrogenase (EC 1.4.1.2) Activity

Principle

Glutamate dehydrogenase catalyses the following reaction

$$\text{Glutamate} + NAD^+ \xrightarrow{\text{GDH}} NH_3 + \text{2-oxoglutarate} + NADH + H^+$$

The enzyme causes the oxidation of reduced coenzymes NADH or NADPH. These reduced coenzymes absorb light at 340 nm. With their use in the reaction with time, change in absorbance is recorded.

Glassware and Equipments

Balance, test tubes, test tube stand, pipettes, measuring cylinder, beakers, volumetric flasks, uv–vis spectrophotometer, centrifuge, centrifuge tubes, pestle and mortar, dialysis tubing

Reagents

1. 0.1 M phosphate buffer (pH 7.5) - See annexure Table 1.7.
2. 0.2 M Tris-HCl buffer (pH 8.3) - Dissolve 24.2 g Tris in 800 ml of water, set pH to 8.3 with HCl and make volume to 1*l*.
3. 0.2 M α-ketoglutaric acid - Dissolve 2.92 g α- ketoglutaric acid in distilled water and make final volume to 100 ml.
4. 1.5 M ammonium sulphate - Dissolve 19.81 g ammonium sulphate in distilled water and make final volume to 100 ml.
5. 4 mM NADH - Dissolve 28.37 mg of NADH in distilled water and make volume to 10 ml.

Procedure

1. Macerate 500 mg of fresh tissue (nodules/ seeds/ pods/ roots) in 10 ml of 0.1 M phosphate buffer (pH 7.5) containing 5 mM cysteine and 300 mg PVP.
2. Centrifuge the contents at 10,000 x g for 10 minutes.
3. Allow the proteins in the supernatant to precipitate using 70% ammonium sulphate concentration and keep it overnight.

4. Next day pellet the proteins by centrifugation at 15,000 x g for 20 minutes.
5. Dissolve the pellet so obtained in 0.1 M phosphate buffer (pH 7.5) and dialyse for 36 hours.
6. Keep this extract for enzyme assay.
7. Take 1.0 ml of 0.2 M Tris HCl buffer (pH 8.3), 0.2 ml of 0.2 M α-ketoglutaric acid, 0.2 ml of 1.5 M ammonium sulphate, 0.2 ml of 4 mM NADH and 0.1 ml enzyme solution.
8. Record the decrease in absorbance at 340 nm every 30 seconds for 3 minutes at 30°C.
9. Express the enzyme activity as μmoles of NADH oxidized min^{-1} mg^{-1} protein or min^{-1} g^{-1} fresh weight of tissue.

Note

- For taking absorbance in UV range use quartz cuvette.

8.8. Determination of Nitrate Reductase (EC 1.6.6.1) Activity

Principle

Nitrate reductase mediates the reduction of nitrate (NO_3^-) to nitrite (NO_2^-) by the reaction

$$NO_3^- + 2H^+ + 2e^- \rightarrow NO_2^- + H_2O$$

Enzyme activity is measured by colorimetric determination of nitrite produced in the reaction.

Glassware and Equipments

Nitrogen gas, screwcapped vial, shaker, test tubes, test tube stand, pipettes, measuring cylinder, beakers, volumetric flasks, uv–vis spectrophotometer

Reagents

1. 0.1 M phosphate buffer (pH 7.5) - See annexure Table 1.7.
2. n–Propanol
3. KNO_3
4. Colouring reagent (1% sulphanilamide and 0.02% N-1-naphthyl ethylene diamine dihydrochloride in 1N HCl) - Dissolve 1 g sulphanilamide and 20 mg N-1-naphthyl ethylene diamine dihydrochloride in 1 N HCl and make final volume to 100 ml.

Procedure

1. Suspend 200 mg of fresh tissue in a screw cap vial containing 5 ml of 0.1 M phosphate buffer (pH 7.5), 0.05 ml of n-propanol and 0.25 g KNO_3.
2. Pass nitrogen gas (N_2) through the incubation media to make the atmosphere of vial free from oxygen. Seal the vials and incubate in a metabolic shaker at 30°C in dark for 90 minutes.
3. Take 1 ml aliquot from above vial containing released nitrite (NO_2^-) and add 1 ml of colouring reagent. Make the total volume to 10 ml with distilled water.
4. After 20 minutes, record the absorbance at 540 nm.
5. Prepare the standard curve using KNO_2 in the range 0-10 μg.
6. Express nitrate reductase activity as μmoles of NO_2^- formed h^{-1} g^{-1} fresh weight of tissue.

8.9. Determination of Peroxidase (EC 1.11.1.7) Activity

Principle

Peroxidase catalyses oxidation of many organic compounds AH_2 (amines eg. benzidine; phenols eg. pyrogallol, guaiacol, hydroquinones) by H_2O_2.

$$AH_2 + H_2O_2 \rightarrow 2H_2O + A$$

When guaiacol is used as a substrate it gets converted to oxidised form or it is dehydrogenated. Rate of formation of dehydrogenated product is a measure of the rate of enzyme activity.

Glassware and Equipments

Test tubes, test tube stand, pipettes, measuring cylinder, beakers, volumetric flasks, uv–visible spectrophotometer, centrifuge, centrifuge tubes, pestle and mortar.

Reagents

1. 0.01 M and 0.05M phosphate buffer (pH 7.0) - See annexure Table 1.7. Dilute and 0.05 M buffer, respectively 0.1 M buffer 10 times and dilute 5 times to get 0.01 M buffer, before use.
2. 20 mM guaiacol - Dissolve 240 mg guaiacol in distilled water and make volume to 100 ml.

3. 12.3 mM or 0.042% H_2O_2 - Dilute 0.14 ml of 30% H_2O_2 to 100 ml with distilled water. The extinction of this should be 0.485 at 240 nm.

Procedure

1. Macerate fresh tissue (0.5 g) in 5 ml 0.05 M phosphate buffer (pH 7.0) containing 1% polyvinylpyrrolidone.
2. Centrifuge the contents at 15,000 x g for 30 minutes.
3. Collect the supernatant for enzyme assay.
4. The reaction mixture contains 3 ml of 0.01 M potassium phosphate buffer (pH 7.0), 0.05 ml of guaiacol and 100 μl enzyme extract.
5. Initiate the reaction by adding 0.03 ml of 12.3 mM H_2O_2.
6. Record the increase in absorbance at 30 seconds interval up to 3 minutes at 470 nm due to the formation of tetraguaicol.
7. Run a blank containing complete reaction mixture without H_2O_2.
8. Express peroxidase activity as the change in absorbance per min per mg soluble protein and calculate activity using extinction coefficient 6.39 $mM^{-1}cm^{-1}$.

8.10. Determination of Ascorbate oxidase (EC 1.10.3.3) Activity

Principle

The enzyme catalyses the direct oxidation of ascorbic acid by molecular oxygen.

$$\text{Ascorbic acid} + \tfrac{1}{2}O_2 \xrightarrow{\text{AAO}} \text{Dehydroascorbic acid} + H_2O$$

Ascorbic acid has absorption maxima at 265 nm. Oxidation of acid causes decrease in its absorption which is measured with time.

Glassware and Equipments

Test tubes, test tube stand, pipettes, measuring cylinder, beakers, volumetric flasks, uv–vis spectrophotometer, centrifuge, centrifuge tubes, pestle and mortar

Reagents

1. 0.02 M and 0.05 M phosphate buffer (pH 7.0) - See annexure Table 1.7

2. 2.5 mM ascorbic acid - Dissolve 44.03 mg ascorbic acid in water and make final volume to 100 ml.
3. 1% polyvinylpyrrolidone - Dissolve 1 g polyvinylpyrrolidone in 0.05 M phosphate buffer and make final volume to 100 ml with buffer.

Procedure

1. Grind fresh tissue (0.5 g) with 5 ml solution containing 0.05 M phosphate buffer (pH 7.0) containing 1% polyvinylpyrrolidone.
2. Centrifuge the homogenate at 15000 x g for 30 min.
3. Collect the supernatant for enzyme assay.
4. Take 1 ml of reaction mixture containing 0.02 M potassium phosphate buffer (pH 7.0) and 2.5 mM ascorbic acid.
5. To it add 10 μl of enzyme extract.
6. Record the decrease in absorbance for 3 min at 265 nm due to ascorbate oxidation and calculate the activity using extinction coefficient of 14 $mM^{-1}cm^{-1}$.

8.11. Determination of β-1, 3-Glucanase (EC 3.2.1.39) Activity

Principle

β-1, 3-glucanase catalyzes hydrolytic cleavage of β-1,3-D-glucosidic linkages in β-1,3-glucans, present in the cell wall of most fungal pathogens to release oligosaccharides. The products of enzymatic reaction are detected by using colouring reagents.

Glassware and Equipments

Test tubes, test tube stand, pipettes, measuring cylinder, beakers, volumetric flasks, spectrophotometer, centrifuge, centrifuge tubes, pestle and mortar, glass capillaries, water bath

Reagents

1. 4% Laminarin -Dissolve 0.4 g of Laminarin in 10 ml of 0.05M of sodium acetate buffer (pH 5.2).
2. 0.05 M of sodium acetate buffer (pH 5.2) - See annexure Table 1.4. Double dilute 0.1 M buffer before use to get 0.05 M buffer.

3. 1% Dinitrosalicylic acid reagent (DNSA) - Dissolve 1 g DNS, 0.2 g phenol and 0.05 g sodium sulphite and 1 g sodium hydroxide in minimum quantity of water and finally make the volume to 100 ml.
4. 40% potassium sodium tartarate - Dissolve 40 g potassium sodium tartarate in distilled water to make final volume to 100 ml.
5. Glucose standard (1 mg/ ml) - Dissolve 100 mg glucose in minimum quantity of distilled water and make the volume to 100 ml.

Procedure

1. Weigh 1 g fresh sample and grind in pre–chilled pestle and mortar using glass capillaries in 3 ml of ice cold 0.1 M sodium acetate buffer (pH 4.8) containing 15 mM of β-mercaptoethanol.
2. Centrifuge the extract at 10,000 rpm for 15 min at 4°C.
3. Keep the supernatant for enzyme assay.
4. Add 0.2 ml of enzyme and 0.5 ml of 4% laminarin to the assay mixture and incubate at 40°C for 1 h.
5. After incubation add 375 μl of DNSA reagent and heat in boiling water bath for 10 minutes to develop red brown colour.
6. Add 1 ml of 40% potassium sodium tartarate to stabilize the colour.
7. Read the absorbance immediately at 575 nm.
8. Enzyme activity is expressed as mg of glucose formed/h at 40°C. The concentration of glucose is read from standard curve prepared by using glucose in range of 0.2–1.0 mg.

8.12. Determination of Phenylalanine Ammonia Lyase (EC 4.3.1.5) Activity

Principle

Phenylalanine ammonia lyase catalyses deamination of phenylalanine to trans–cinnamic acid. Appearance of trans–cinnamic acid is read at 290 nm.

Glassware and Equipments

Test tubes, test tube stand, pipettes, measuring cylinder, beakers, volumetric flasks, uv–vis spectrophotometer, centrifuge, centrifuge tubes, pestle and morar

Reagents

1. 0.03 M phenylalanine in 0.05 M sodium borate buffer (pH 8.8) - Dissolve 4.96 g phenylalanine in sodium borate buffer to make final volume to 1000 ml.
2. 0.05 M sodium borate buffer (pH 8.8) - Refer Annexure Table 2.0 Double dilute 0.1 M borate buffer to get 0.05 M buffer.

Procedure

1. Weigh 150 mg sample and grind in pre-chilled pestle and mortar using glass capillaries in 3.5 ml of ice cold 0.1 M phosphate buffer (pH 7.5) containing 5 mM of β-mercaptoethanol.
2. Centrifuge the extract at 10,000 rpm for 25 min at 4°C.
3. Keep the supernatant for enzyme assay.
4. Take 2.7 ml of 0.03 M L-phenylalanine prepared in 0.05 M sodium borate buffer (pH 8.8) and add 0.3 ml of enzyme extract.
5. Incubate the reaction mixture at 37°C for 1 hr.
6. Stop the reaction with 0.3 ml of 1 N HCl.
7. Record the appearance of t-cinnamic acid resulting from the deamination of L-phenylalanine at 290 nm.
8. Reaction mixture without incubation for 1 h serves as control.
9. Enzyme activity is expressed as μg t-cinnamic acid formed h^{-1} g^{-1} fresh weight and μg t-cinnamic acid formed h^{-1} mg^{-1} protein. The concentration of t-cinnamic acid is read from standard curve prepared for t-cinnamic acid in the range of 5-40 μg.

9 SECONDARY METABOLITES

Apart from primary metabolites such as proteins, carbohydrates, lipids, chlorophylls, cytochromes and various metabolic intermediates, plants also contain a large variety of secondary metabolites. These metabolites have no direct metabolic functions. But plants often produce these for protection against pathogens and herbivores. They include alkaloids, isoprenoids or terpenoids, phenolic compounds etc. Alkaloids are a group of secondary metabolites that are synthesized from amino acids and contain one or more N atoms in their heterocyclic structure. They are basic in nature and are stored in vacuole, which is acidic in nature. More than 10,000 alkaloids of different structures are now known. Capsaicin, found in chillies is a protoalkaloid. Nicotine (tobacco), cocaine (*Atropa belladonna*), quinine (*Chinchuna officinalis*), and caffeine (coffee) are a few other examples.

Isoprenoids also called as terpenoids and are present in all living organisms, but with unusual diversity in plants. A number of cyclic compounds containing 10, 15, 20 or more C atoms have been isolated from turpentine oil. Such substances have been found in other plants and given the collective name terpenes. Isoprene is the basic unit of terpenes.

Carotenoids are aterpene subclass which consists of bright yellow, orange and red plant pigments found in vegetables such as tomatoes, oranges, spinach, red palm oil etc.

Plant phenolics, with more than 8,000 identified compounds, constitute one of the largest and most ubiquitous groups of secondary metabolites that constitute one of the common and widespread groups of substances in plants. The term phenolic or polyphenol can be characterized structurally by the presence of one or more phenol rings, and two or more hydroxyl groups linked directly to the aromatic rings. They can be classified into several structural classes - phenolic acids, flavonoids, anthocyanins, stilbenes, lignans, lignins, tannins etc.

In plants, phenolics have many diverse functions: pigmentation, growth, reproduction, resistance to pathogens, antimicrobial and antifungal properties, protection from UV radiation, heavy metal chelation, and antioxidant protection from free radicals generated during photosynthesis.

Gossypol is a polyphenolic yellow pigment found naturally throughout the cotton plant.In the seeds, it is found in small pigment glands. when the seed kernels are flaked and heated as part of oil extraction process, these small glands are ruptured and the gossypol is released. Members of family Brassicaceae, cruciferous vegetables including rapeseed-mustard, cabbage, cauliflower, broccoli, kale, Brussels sprouts etc are the dietary source of glucosinolates. These compounds characterize this family and are responsible for the characteristic pungent flavour of mustard and mustard products. Glucosinolates a large group of sulphur–containing glucosides, are secondary plant metabolites derived from different amino acids such as methionine, phenylalanine and tryptophan. Structurally, they are sulphonated oxime thioesters of β-D-glucose. The moiety attached to thiohydroximate or side chain (R) varies and results in 130 types of glucosinolates. R chain could be aliphatic, cyclic or heterocyclic.These compounds remain intact unless brought into contact with the enzyme myrosinase (thioglucosidase, EC 3.2.3.1) by pests, food processing, or chewing. Intact compounds are nontoxic while their hydrolysis products are considered harmful to health of non ruminants such as poultry and pigs as they interfere with thyroid function and cause liver damage.

Qualitative Analysis

9.1. Extraction of Metabolites

Take 5 g of finely ground dried powder of given plant sample in an Erlenmeyer flask. Add 25 ml of distilled water to it. Heat the contents of flask at 50–60°C on a water bath for 30 min. Cool and filter the contents through Whatman filter paper no.1.Centrifuge the filtrate at 2500 rpm for 15 min and decant the supernatant and keep it at 5°C for further use.

9.2. Test for Steroids

Principle

H_2SO_4 when added to sterols, gives a green coloured chromophore.

Glassware and Equipments

Test tubes, test tube stand, pipettes, measuring cylinder

Reagents

1. Chloroform
2. Concentrated sulphuric acid

Procedure

Take 1 ml of extract and dissolve it in 10 ml of chloroform. Add 10 ml of concentrated sulphuric acid (H_2SO_4) by the side of tube. The upper layer becomes red and lower sulphuric acid layer becomes yellow with green florescence.

9.3. Test for Tannins

Principle

Tannins are complex phenolic compounds, which can form soluble and insoluble complexes with proteins and as a result reduce the availability of the proteins. They form yellow precipitates with lead acetate.

Glassware and Equipments

Test tubes, test tube stand, pipettes, dropper

Reagents

1. Lead acetate (1%): Dissolve 1 g of lead acetate in water and make the volume to 100 ml.

Procedure

Take 2 ml of extract and add a few drops of 1% lead acetate to it. Formation of yellow precipitates shows the presence of tannins.

9.4. Test for Saponins

Principle

These compounds make foam on shaking with water. So formation of foam is the basis for their detection.

Glassware and Equipments

Test tubes, test tube stand, pipettes, graduated tube

Reagents

1. Distilled water

Procedure

Take 5 ml of extract in a graduated tube. Add 20 ml of distilled water to it. Agitate the contents of the flask for 30 sec to a few min. Formation of foam indicates presence of saponins. If level of foam is 5 mm or less- negligible saponins; 5–9 mm - low content of saponins; 10–14 mm - medium content of saponins foam; 15 mm or more-sample contains high amount of saponins.

9.5. Test for Terpenoids

Principle

H_2SO_4 when added to terpenoids gives a green–coloured chromophore in the presence of acetic anhydride. Steroids also give this test positive since they are also derived from isoprene.

Glassware and Equipments

Test tubes, test tube stand, pipettes

Reagents

1. Acetic anhydride
2. Concentrated sulphuric acid

Procedure

Take 2 ml of extract and to it add 2 ml of acetic anhydride and 2 ml of concentrated sulphuric acid (H_2SO_4). Formation of bluish–green coloured ring shows the presence of terpenoids.

Quantitative Analysis

9.6. Determination of Total Phenol Content

Principle

Phenols react with phosphomolybdic acid of Folin Ciocalteau reagent in alkaline medium to produce blue–coloured complex. The blue colour formed is read at 760 nm.

Glassware and Equipments

uv–vis spectrophotometer, test tubes, test tube stand, pipettes, water bath, ice bath, centrifuge, centrifuge tubes, pestle and mortar

Reagents

1. Folin Ciocalteau reagent (2N): Use as it is.
2. Gallic acid (1 mg/1 ml): Dissolve 100 mg of gallic acid in ethanol and make the volume to 100 ml.
3. Sodium carbonate (20%): Dissolve 20 g of sodium carbonate in water and make the volume to 100 ml.

Procedure

1. Macerate 100 mg of sample in pestle and mortar rising 80% methanol.
2. Centrifuge the contents at 5000 rpm for 10 minutes.
3. Rextract the residue with 80% ethanol.
4. Pool the supernatants and note the volume of extract.
5. Take 100 μl of sample solution and to it add 7.9 ml of distilled water.
6. Add 500 μl of Folin Ciocalteau reagent.
7. Incubate the tubes for 10 min at room temperature.
8. Add 1.5 ml of 20% sodium carbonate solution.
9. Similarly, prepare a water blank without adding the sample.
10. Heat the tubes at 40°C for 20 min and cool the tubes in an ice bath immediately.
11. Read absorbance at 760 nm.
13. Prepare a standard curve using gallic acid in the range of 100–500 μg and express the results as gallic acid equivalents i.e., GAE (mg/g GAE).

9.7. Determination of Flavonoids

Principle

Flavonoids form complex with aluminum nitrate and produce blue colour. Intensity of the blue colour developed is measured at 415 nm.

Glassware and Equipments

Balance, uv–vis spectrophotometer, test tubes, test tube stand, pipettes

Reagents

1. Ethanol
2. Aluminum nitrate [Al $(NO_3)_3$] (10%): Dissolve 10 g of aluminum nitrate in water and make the volume to 100 ml.
3. Potassium acetate (CH_3COOK) (1M): Dissolve 1 g mol wt of potassium acetate in water and make the volume to 1*l*.
4. Quercetin (1 mg/ 10 ml): Dissolve 5 mg in ethanol and make the volume to 50 ml.

Procedure

1. Take 100 μl of extract of the sample as prepared in 9.6 Ex. in a tube.
2. Add 1.5 ml of ethanol and mix thoroughly.
3. Add 100 μl of 10% aluminum nitrate solution.
4. Add 100 μl of 1 M potassium acetate and 2 ml distilled water.
5. Similarly, prepare a water blank without the sample.
6. Keep the tubes at room temperature for 40 min.
7. Read absorbance at 415 nm.
8. Prepare a standard curve using quercetin in the range of 2–10 μg and express the results as quercetin equivalents (μg/g).

9.8. Determination of Tannins

Principle

Tannins reduce phosphotungstomolybdic acid in alkaline solution to produce a blue colour, the intensity of which is proportional to the amount of tannins, and is measured at 725 nm.

Glassware and Equipments

Balance, test tubes, test tube stand, pipettes, measuring cylinder, beakers, volumetric flasks, spectrophotometer, pestle and mortar, electrical shoker, conical flask

Reagents

1. Diethyl ether containing 1% acetic acid: Mix 1 ml acetic acid with 99 ml diethyl ether.
2. 70% aqueous acetone - Take 70 ml acetone and add 30 ml distilled water.
3. Folin-Denis reagent - Dissolve 100 mg sodium tungstate and 20 mg phosphomolybdic acid in 750 ml distilled water in a flask. Add 50 ml phosphoric acid. Reflux the mixture for 2 h and make up to 1*l* with water and store in a brown bottle.
4. 20% Sodium carbonate solution - Dissolve 200 g sodium carbonate in 1*l* of distilled water at 70–80°C. Filter through Whatman filter paper no. 1 after keeping the solution overnight.
5. Standard tannic acid (0.5 mg/ml) - Dissolve 5 mg tannic acid in distilled water and make the volume to 10 ml.

Procedure

1. Wash the plant material thoroughly and dry it under shade.
2. Grind the dry sample and sieve it through 1 mm diameter mesh.
3. Weigh 400 mg and transfer to a conical flask with 40 ml diethyl ether containing 1% acetic acid (v/v) and mix to remove the pigments.
4. Discard the supernatant carefully after 5 min and add 20 ml of 70% aqueous acetone.
5. Cover the flask with cotton plug followed by aluminium foil and keep in an electrical shaker for 2 h for extraction.
6. Filter it through Whatman filter paper no. 1 and keep the sample in refrigerator at 4°C until analysis.
7. Take 0.1 ml of extract and make the volume to 1 ml with distilled water.
8. Similarly prepare the standards from stock solution of tannic acid (0.5 mg/ml) taking 0, 10, 20, 30, 40 and 50 µl in test tubes and make the volume to 1.0 ml with distilled water. This gives a tannic acid concentration of 0, 5, 10, 15, 20 and 25 µg, respectively.

9. Add 0.5 ml Folin–Denis reagent and 2.5 ml 20% sodium carbonate. Mix thoroughly and after 40 minutes take reading at 725 nm on spectrophotometer.
10. Tannic acid content is calculated as mg of tannic acid/ g dry weight of sample.

9.9. Determination of Gossypol

Principle

Gossypol reacts with phloroglucinol under acidic conditions to produce a reddish brown product, which is measured at 550 nm.

Glassware and Equipments

Balance, test tubes, test tube stand, pipettes, measuring cylinder, beakers, volumetric flasks, spectrophotometer, Separatory funnel, blade water bath, pH strips/pH meter

Reagents

1. 95% Ethanol - Take 95 ml of ethanol and add 5 ml of distilled water.
2. Phloroglucinol reagent - Dissolve 5 g phloroglucinol in 100 ml of 80% ethanol (80 ml ethanol mix with 20 ml of distilled water).
3. 40% Ethanol - Take 40 ml of ethanol and add 60 ml of distilled water.
4. Concentrated HCl

Procedure

1. Weigh 5 g of fresh tissue, cut it into small pieces.
2. Boil the pieces for 5 minutes in 20 ml of 95% ethanol.
3. Filter the extract and repeat the boiling process once again and combine the extracts.
4. Dilute the extract with 40% ethanol and adjust the pH to 3.0 with 1N HCl and transfer to a separatory funnel.
5. Add 1.5 volumes of diethyl ether at 10°C. Keep the ether phase and wash it with distilled water twice.
6. Evaporate the ether extract *in vacuum* to dryness and redissolve the residue in a known volume of 95% ethanol.

7. Pipette out different aliquots (1, 2 ml) of the gossypol extract in test tubes.
8. Add 0.5 ml of phloroglucinol reagent followed by 1 ml of conc. HCl to each tube.
9. Incubate for 30 minutes at room temperature with occasional shaking.
10. Make the volume to 10 ml with 80% ethanol and take the absorbance at 550 nm.
11. Prepare the standard curve with gossypol acetate.

9.10. Determination of Capsaicin

Principle

Capsaicin, a protoalkaloid responsible for pungency of chillies, contain phenolic groups, which reduce phosphomolybdic acid to form a blue–coloured complex that is read at 650 nm.

Glassware and Equipments

Balance, test tubes, test tube stand, pipettes, measuring cylinder, beakers, volumetric flasks, uv–vis spectrophotometer, centrifuge, centrifuge tubes, water bath

Reagents

1. 0.4% Sodium hydroxide - Weigh 4 g sodium hydroxide and dissolve in distilled water and make the volume to 1*l*.
2. 3% Phosphomolybdic acid - Weigh 3 g phosphomolybdic acid and dissolve in distilled water and make the volume to 100 ml.
3. Dry acetone - Add 25 g anhydrous sodium sulphate to 500 ml acetone, one day before use.
4. Capsaicin standard (200 μg/ ml) - Weigh 10 mg capsaicin and dissolve in 0.4% sodium hydroxide solution and finally make the volume to 10 ml to make the stock standard of 1 mg/ml. Dilute stock standard 5 times to get working standard of 200 μg/ml.

Procedure

1. Weigh 0.5 g dry chilli powder and transfer to a volumetric flask.

2. Add 10 ml of dry acetone to it and keep on shaker for 3 h.
3. Centrifuge at 10000 rpm for 10 minutes.
4. Take 1 ml of supernatant into a test tube and evaporate to dryness in a hot water bath.
5. Dissolve the residue in 5 ml of 0.4% sodium hydroxide. Add 3 ml of phosphotungstic acid. Mix the contents and keep the tubes at room temperature for 1 h.
6. Filter the solution into centrifuge tubes to remove any floating debris and centrifuge at 5000 rpm for 15 minutes.
7. Transfer the clear supernatant into cuvette and read at 650 nm.
8. Run the reagent blank along with test samples.
9. For standard pipette out 0.2, 0.4, 0.6, 0.8 and 1 ml of working standard solution and proceed as above.

9.11. Determination of Lycopene

Principle

Lycopene is red–coloured pigment found in tomato, having formula $C_{40}H_{56}$. The carotenoids in the sample are extracted in acetone and then transferred to petroleum ether. 1M lycopene in petroleum ether (40–60°C) gives an absorbance of 17.2×10^4 at 503 nm. Therefore 1 unit of absorbance will be given by 3.12 μg/ml lycopene.

Glassware and Equipments

Balance, conical flasks, pipettes, measuring cylinder, beakers, volumetric flasks, spectrophotometer, separatory funnel, pestle and mortar

Reagents

1. Acetone
2. Petroleum ether (40–60°C)
3. Anhydrous sodium sulphate
4. 5% Sodium sulphate - Dissolve 5 g sodium sulphate in distilled water and make the volume to 100 ml.

Procedure

1. Weigh 10 g of tomato pulp and extract it with acetone in pestle and mortar until residue is colourless.

2. Pool acetone extracts and transfer to separatory funnel containing 20 ml petroleum ether and mix gently.
3. Add 20 ml of 5% sodium sulphate solution and shake gently. Add 20 ml more of petroleum ether and mix and allow for layer separation. Red colour will appear in the ether layer.
4. Take ether layer in a stoppered container. Re–extract the aqueous layer with 20 ml lots of petroleum ether till the aqueous layer is colourless.
5. Pool the ether extracts and give washing with small amount of distilled water.
6. Transfer the extract in brown bottle containing 10 g anhydrous sodium sulphate and keep for 1 h.
7. Filter the extract into 100 ml volumetric flask through cotton wool plug. Give washings to sodium sulphate slurry with petroleum ether until the slurry is colourless. Also filter the washings through cotton plug into the flask. Make the volume to 100 ml.
8. Measure the absorbance at 503 nm using petroleum ether as blank.
9. Calculate the amount of lycopene in mg/100 g sample by multiplying with factor 3.12 with absorbance of sample at 503 nm.

9.12. Determination of Curcumin

Principle

Curcumin is extracted from tissue by refluxing in absolute alcohol and estimated spectrophotometrically at 425 nm.

Glassware and Equipments

Balance, test tubes, test tube stand, pipettes, measuring cylinder, beakers, flasks with volumetric flasks, uv–vis spectrophotometer, water bath, condensess

Reagents

1. Absolute ethanol.

Procedure

1 Weigh 0.5 g of dry turmeric powder and dissolve it in 250 ml of absolute ethanol.

2. Reflux the contents for 4 h.
3. Decant the extract into volumetric flask and make up the volume to 50 ml.
4. Dilute the extract 10 times with absolute ethanol and measure the intensity of the colour at 425 nm.
5. Calculate the curcumin content as g/100 g tissue by taking into account the fact that 0.42 absorbance = 2.5 mg curcumin.

9.13. Determination of Anthocyanins

Principle

Anthocyanins are colouring pigments in plants responsible for nearly all colours in different tissues. They are extracted in alcohol. The ethanol extract is treated with methanolic HCl followed by reaction with colouring agent. The developed colour is read at 525 nm.

Glassware and Equipments

Balance, test tubes, test tube stand, pipettes, measuring cylinder, beakers, volumetric flasks, spectrophotometer, centrifuge, centrifuge tubes, pestle and mortar

Reagents

1. 95% Ethanol - Take 95 ml of ethanol and add 5 ml of distilled water.
2. 0.5 N HCl in 80% methanol - Pipette 43 ml concentrated HCl into 1000 ml volumetric flask and make the volume to100 ml with 80% methanol (80 ml methanol and 20 ml distilled water).
3. Anthocyanin / Colouring reagent - Mix 1 ml of 30% H_2O_2 with 9 ml of methanolic HCl.

Procedure

1. Weigh 0.5 g of fresh plant tissue and grind in 95% ethanol. Centrifuge the extract. Collect the supernatant.
2. Pipette 1 ml of supernatant into a test tube and add 3 ml of HCl in aqueous methanol followed by addition of 1 ml anthocyanin reagent.
3. Incubate in dark for 15 minutes and take the absorbance at 525 nm.

4. Calculate the anthocyanin content in sample by using cyanidin - 3- glucoside as a stanard (molar extinction coefficient of 26,900 L cm^{-1} mol^{-1}; N. wt. 449.2 g mol^{-1}).

9.14. Determination of Total Glucosinolates

Principle

A coloured complex is formed between glucosinolates and tetrachloropalladate. A colour intensification from light to dark brown is observed on incubation with $[Pd(Cl_4)]^{2-}$ at 70°C. The color intensity is measured at 405 nm using ELISA reader.

Glassware and Equipments

Balance, micropipettes, measuring cylinder, beakers, volumetric flasks, microscan ELISA reader, ELISA plate, pestle and mortar, oven water bath

Reagents

1. Methanol (80%) - Mix 80 ml methanol with 20 ml distilled water.
2. Sodium tetrachloro palladate (0.002M or 2 mM) solution - Dissolve 58.8 mg of sodium tetrachloro palladate, add 177 μl of conc. HCl and make final volume to 100 ml.

Procedure

1. Crush 200 mg of oven dried seed in a mortar pestle and transfer it to a screw capped tube.
2. Add 300μl of 70% methanol and keep it in water bath (70°C) for 5 minutes.
3. Cool at room temperature and add 2 ml of distilled water.
4. Keep the tubes in water bath (70°C) for 15 minutes.
5. Cool at room temperature and centrifuge it at 5000 rpm for 5 minutes.
6. Transfer 5 μl from supernatant to ELISA plate and add 300 μl of 0.002 M sodium tetrachloropalladate solution to each sample in ELISA plate.
7. Keep the plate in oven for 70°C for 30 minutes.
8. Record intensity of colour using ELISA reader at 405 nm.

Calculation

Total glucosinolate concentration

$= (A_{405} - 0.0465)*82$, for $A_{405} < 0.8$ (μ moles /g meal)

Total glucosinolate concentration

$= (A_{405} - 0.169) *115.74$, for $A_{405} > 0.8$ (μ moles /g meal)

Where A_{405} is absorbance at 405 nm.

10 ELECTROPHORESIS

Electrophoresis (electro refer to the energy of electricity and phoresis, from the greek verb phoros, means to carry across) is a technique for separating, or resolving, charged molecules (such as amino acid, peptides, proteins, nucleotides, and nucleic acids) in a mixture under the influence of an applied electric field. Charged molecules in an electric fields move or migrate, at a rate determined by their charge/mass ratio. According to the laws of electrophoresis an ion with charge 'Q' in an electric field of strength 'E' will experience a electric force

$$(F_{electric}) = QE$$

The resulting migration of the charged molecules through the solution is opposed by a frictional force

$$F_{frictioional} = Vf$$

Where V is the rate of migration of ion and f is its frictional coefficient. It depends on the size, shape and viscosity of the solution. In constant electric field, the force on ion balance each other i.e. QE=Vf so that each ion moves with a constant characteristic velocity. An ion's electrophoretic mobility (μ) is given as μ=V/E= Q/f

According to equation, if two molecules have the same mass and shape, the one with greater net charge will move faster toward an electrode. Electrophoresis is of two types- moving boundary electrophoresis and zone electrophoresis. In zone electrophoresis sample is constrained to move in a solid support such as filter paper (paper electrophoresis) or a gel (gel electrophoresis).

10.1. Sodium Dodecyl Sulphate - Polyacrylamide Gel Electrophoresis (SDS PAGE)

Principle

SDS PAGE is the most commonly used method for separation of proteins from a mixture.SDS is a negatively–charged detergent. When proteins are boiled in the presence of mercaptoethanol and SDS, proteins get reduced and bind with SDS getting its negative charge. So proteins are separated only on the basis of their molecular weight. By using standard proteins of known molecular weight, molecular weight of unknown proteins can be determined. Mobility of protein in the gel is noted with respect to the tracking dye, bromophenol blue and is mentioned as Rf (Figs. 13-15).

Rf = Distance travelled by protein/Distance travelled by tracking dye

Glassware and Equipments

Balance, micropipettes, beakers, reagent bottles, vertical slab gel electrophoresis apparatus (two glass plates - one notched and one plain, spacers, combs, gel casting unit, electrophoresis tank unit, electrodes), power supply

Reagents

1. Resolving gel buffer (Solution A1)

1.87 M Tris HCl buffer pH 8.8 - Dissolve 22.69 g Tris in distilled water and make final volume to 100 ml and set the pH using HCl.

2. Stacking gel buffer (Solution A2)

0.6 M Tris HCl buffer pH 6.9 - Dissolve 7.26 g Tris in distilled water and make final volume to 100 ml and set the pH using HCl.

3. 10% SDS (Solution A3)

Dissolve 10 g SDS in distilled water and make volume to 100 ml.

4. 0.5% APS (Solution A4)

Dissolve 0.5 g of Ammonium sulphate in distilled water and make volume to 100 ml. Prepare it freshly before use.

5. Stock Protein Extraction (Solution A5)

SDS 2 g, Tyrosine 1 g

Mix both with 10.4 ml of 0.6 M Tris HCl buffer pH 6.9 (A2) and add 7.9 ml of distilled water and 10 ml of glycerol. Boil the mixture for 15 min to make uniform solution.

6. Working Protein Extraction (SolutionA6)

Take 4.25 ml of solution A5, add 0.75 ml β–mercaptoethanol and make final volume to 10 ml.

7. Electrode buffer A7

SDS - 1.0 g
Tris - 3.0 g
Glycine - 14.1 g, final volume 1 l with distilled water.

8. Fixing (Solution A8)

TCA - 150 g, final volume 1l with distilled water.

9. Staining (Solution A9)

Coomassie brilliant blue-1 g, final volume 100 ml with distilled water.

Mix 100 ml of this solution with 100 ml of fixing solution A8 to form staining solution.

10. Destain solvent mixture A10

Chloroform - 200 ml
Methanol - 100 ml
Acetone- 100 ml

11. Acrylamide (30%) for resolving gel A11

Acrylamide -75 g
Bis-acrylamide -1 g, final volume - 250 ml with distilled water.

12. Acrylamide (30%) for stacking gel A12

Acrylamide - 75 g
Bis-acrylamide - 2 g
Final volume - 250 ml with distilled water.

13. Destaining (Solution A13)

Methanol - 250 ml
Acetic acid - 70 ml, final volume 1 l with distilled water.

14. Gel loading dye

Bromophenol blue -1% 10 mg/ml

15. Composition of resolving & stacking gel solution for SDS PAGE

Resolving gel

Tris buffer (A1) - 6 ml

Distilled water - 3.7 ml

30% acrylamide (A11) – 10 ml

APS (A4) - 0.4 ml

SDS - 0.4 ml

TEMED - 0.04ml (40 μl)

Stacking gel

Tris buffer (A2) - 1.5 ml

Distilled water - 6.0 ml

30% acrylamide (A11) - 2.0 ml

APS (A4) - 0.4 ml

SDS- 0.3 ml

TEMED - 0.04 ml (40 μl)

Procedure

1. Assemble a vertical slab gel unit in casting mode using 1.5 mm spacers.
2. Pour running gel solution into gel sandwich.
3. Overlay the gel with water and allow the gel to polymerize for one hour.
4. Once the gel is polymerized, remove the water, pour the stacking gel solution to insert the combs.
5. Carefully insert the comb while avoiding air bubbles.
6. When the gel gets polymerizes, remove the comb and wash the wells thoroughly with distilled water and then with small amount of electrophoresis buffer to remove any unpolymerized gel and gel debris.
7. Fill electrophoresis buffer in lower tank.
8. Introduce 35μl of each sample into each well.
9. Attach the upper buffer chamber with gel sandwich and lower into the cell.

10. Fill upper tank with electrophoresis buffer.
11. Connect the assembly with cooling device (optional).
12. Connect the electrodes of unit to the power supply unit to supply constant current of 35mA.
13. Disconnect the power supply when the tracking dye reaches at the bottom of the gel.
14. Carefully remove the gel from the apparatus.
15. Put it in the staining solution.
16. Destain the gel using the destaining solution.
17. Take photograph of the gel and analyze it using Gel Documentation System and Software.

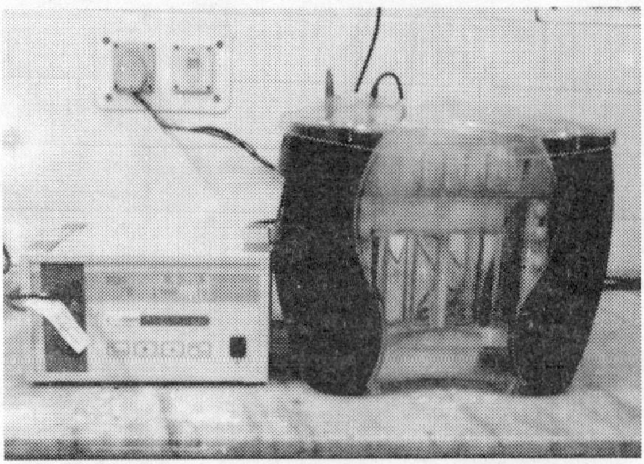

Fig. 13. SDS PAGE apparatus

Fig. 14. Gel casting unit

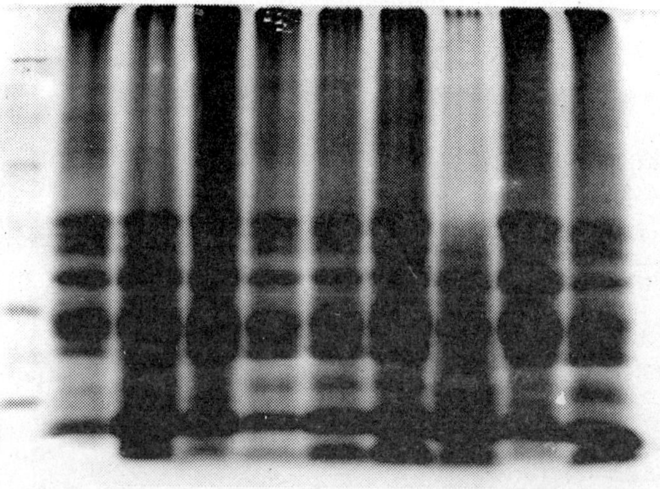

Fig. 15 SDS PAGE showing bands of different proteins

10.2. Agarose Gel Electrophoresis

Principle

DNA samples are separated on agarose gels. DNA fragments move from cathode (DNA is negatively charged) to anode with the same mobility under an applied electric field as the charge per unit length is same. So, the DNA fragments are separated on the basis of their size. Smaller fragments will move faster while bigger fragments will move slowly. DNA fragments are visualized using ethidium bromide dye, which binds to DNA by insertion between stacked base pairs, and gives orange or red fluorescence under UV light (Fig. 16).

Glassware and Equipments

Balance, micropipettes, conical flask, beaker, horizontal slab gel apparatus (gel casting unit, comb, electrophoresis tank), power supply, UV transilluminator

Reagents

1. Agarose: Dissolve 1 g agarose in 100 ml of TAE buffer and heat the solution to get clear solution. Allow it to cool at 60°C and add ethidium bromide.
2. Ethidium bromide (0.5 µg/ml): Prepare a stock solution of 10 mg/ml in distilled water.

3. Tracking dyes: Xylene cynol and bromophenol blue

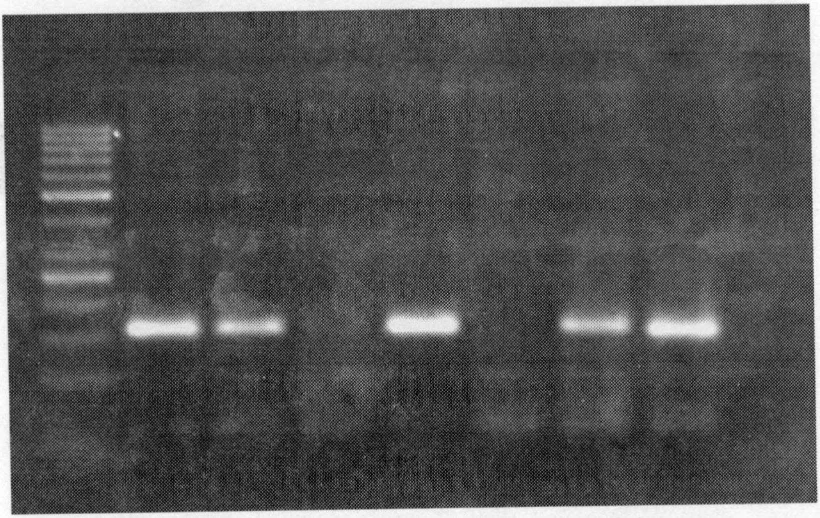

Fig. 16 Agarose Gel electrophoresis showing DNA bands

Procedure

1. Set the gel casting unit, and place the comb in the casting unit in such a way that the comb is about 2 cm away from the cathode.
2. Pour the agarose solution having ethidium bromide in the gel casting unit.
3. Keep the unit undisturbed and allow the gel to solidify (45 minutes).
4. Gently remove the comb from the gel.
5. Pour electrophoresis buffer in the tank unit.
6. Place the gel within the electrophoresis tank unit and connect the power supply. While connecting the electrode care should be taken that negative electrode be attached to the well side of the gel.
7. Load the samples in the wells in a particular sequence.
8. Start the power supply after loading the samples. Set the voltage at 70 V.
9. Allow the gel to run till the tracking dye reaches ¾ th of the gel.

10. Switch off the power supply and disconnect the cords from the respective electrodes.
11. Remove the gel from the tank and place it in UV transilluminator to visualize the DNA bands.

Note

- Always wear gloves while handling ethidium bromide; it is highly carcinogenic.
- Use UV protecting goggles during visualization under UV light.

ANNEXURE

Table 1.1. Densities and normalities of commonly used acids

S. No.	Acid	Density	Approximate normality
1.	Acetic	1.05	17
2.	HCl	1.20	11
3.	HNO_3	1.42	16
4.	H_3PO_4	1.70	15
5.	H_2SO_4	1.84	36

Table 1.2 Composition of DEB

Component	Stock solution	Working concentration	Volume used from stock
Tris (pH 8.0)	1 M	100 mM	10 ml
EDTA (pH 8.0)	0.5 M	20 mM	4
NaCl	4 M	1.4 M	35 ml
CTAB	10%	2%	20 ml
β-mercaptoethanol	-	0.2%	200 μl

Table 1.3 Represents Rf values of standard amino acids and standard sugars

S.No.	Name of amino acid	Rf	Name of sugar	Rf
1.	Alanine	0.23	Arabinose	0.21
2.	Arginine	0.20	Fructose	0.20
3.	Aspartate	0.28	Fucose	0.27
4.	Cystine	0.16	Galactose	0.16
5.	Gtutamine	0.11	Glucose	0.18
6.	Gtutamate	0.35	Lactose	0.09
7.	Glycine	0.37	Maltose	0.11
8.	Leucine	0.72	Mannose	0.20

9.	Phenylalanine	0.69	Raffinose	0.05
10.	Proline	0.13	Ribose	0.31
11.	Serine	0.32	Sucrose	0.14
12.	Valine	0.58	Xylose	0.28

Buffers

Buffers are generally prepared in stock solutions and can be diluted before use as per requirements.

Normally from the following tabulated information 0.1 M buffers are obtained and if exception is there the information is given.

1. Acetate buffer

Stock solutions

A: 0.2 M solution of acetic acid (11.15 ml in 1 l)

B: 0.2 M solution of sodium acetate (16.4 g $C_2H_3O_2.Na$ or 27.2 g $C_2H_3O_2.Na.3H_2O$ in 1 l)

X ml of A+ Y ml of B, diluted to a total volume of 100 ml.

Table 1.4 Acetate buffer

X	Y	pH	X	Y	pH
46.3	3.7	3.6	20.0	30.0	4.8
44.0	6.0	3.8	14.8	35.2	5.0
41.0	9.0	4.0	10.5	39.5	5.2
36.8	13.2	4.2	8.8	41.2	5.4
30.5	19.5	4.4	4.8	45.2	5.6
25.5	24.5	4.6			

2. Citrate buffer

Stock solutions

A: 0.1 M solution of citric acid (21.0 g in 1 l)

B: 0.1 M solution of sodium citrate (29.6 g $C_6H_6O_7.\ Na_2.\ 2H_2O$ in 1 l)

X ml of A + Y ml of B, diluted to a total volume of 100 ml.

Table 1.5. Citrate buffer

X	Y	pH	X	Y	pH
46.5	3.5	3.0	23.0	27.0	4.8
43.7	6.3	3.2	20.5	29.5	5.0
40.0	10.0	3.4	18.0	32.0	5.2
37.0	13.0	3.6	16.0	34.0	5.4
35.0	15.0	3.8	13.7	36.3	5.6
33.0	17.0	4.0	11.8	38.2	5.8
31.5	18.5	4.2	9.5	41.5	6.0
28.0	22.0	4.4	7.2	42.8	6.2
25.5	24.5	4.6			

3. Glycine-NaOH buffer

Stock solutions

A: 0.2 M solution of glycine (15.0 g in 1 l)

B: 0.2 M NaOH (8 g in 1 l)

50 ml of A + X ml of B, diluted to a total of 200 ml.

Table 1.6. Glycine – NaOH buffer

X	pH	X	pH
4.0	8.6	22.4	9.6
6.0	8.8	27.2	9.8
8.8	9.0	32.0	10.0
12.0	9.2	38.6	10.4
16.8	9.4	45.5	10.6

4. Phosphate buffer

Stock solutions

A: 0.2 M solution of monobasic sodium phosphate (27.8 g of NaH_2PO_4 in 1 l)

B: 0.2 M solution of dibasic sodium phosphate (53.7 g $Na_2HPO_4.7H_2O$ or 71.7 g $Na_2HPO_4.12\ H_2O$ in 1 l)

X ml of A + Y ml of B, diluted to a total of 200 ml.

Table 1.7. Phosphate buffer

X	Y	pH	X	Y	pH
93.5	6.5	5.7	45.0	55.0	6.9
92.0	8.0	5.8	39.0	61.0	7.0
90.0	10.0	5.9	33.0	67.0	7.1
87.7	12.3	6.0	28.0	72.0	7.2
85.0	15.0	6.1	23.0	77.0	7.3
81.5	18.5	6.2	19.0	81.0	7.4
77.5	22.5	6.3	16.0	84.0	7.5
73.5	26.5	6.4	13.0	87.0	7.6
68.5	31.5	6.5	10.5	90.0	7.7
62.5	37.5	6.6	8.5	91.0	7.8
56.5	43.5	6.7	7.0	93.0	7.9
51.0	49.0	6.8	5.3	94.7	8.0

5. Tris HCl buffer

Stock solutions

A: 0.2 M solution of tris hydroxyl methyl amino methane (24.2 g in 1 l)

B: 0.2 M HCl

50 ml of A + X ml of B, diluted to a total of 100 ml. This will give 0.1 M buffer.

Table 1.8 Tris HCl buffer

X	pH	X	pH
5.0	9.0	26.8	8.0
8.1	8.8	32.8	7.8
12.2	8.6	38.4	7.6
16.5	8.4	41.4	7.4
21.9	8.2	44.2	7.2

6. 1 M Tris HCl

Dissolve 121.1 g of Tris in 800 ml of distilled water. Adjust the pH to desired value by adding concentrated HCl as below:

Table 1.9. Preparation of 1 M Tris HCl

pH	Amount of HCl (ml)
7.4	70
7.6	60
8.0	42

7. Boric acid - borax buffer

Stock solutions

A : 0.2 M boric acid (12.4 g / litre)

B : 0.05 M sodium tetraborate. 10 H_2O (19.05 g / litre) It is equivalent to sodium borate, 0.2 M $NaBO_2$

X ml of A + Y ml of B, diluted to a total of 200 ml.

Table 2.0. Boric acid - borax buffer

X	Y	pH	X	Y	pH
85	15	7.6	45	55	8.6
80	20	7.8	40	60	8.7
70	30	8.0	30	70	8.8
65	35	8.2	20	80	9.0
55	45	8.4	10	90	9.1

Subject Index